The KEY TOPICS Series

Advisors:

Key Topics in Anaesthesia

Key Topics in Obstetrics and Gynaecology

Key Topics in Accident and Emergency Medicine

Key Topics in Paediatrics

Forthcoming titles include:

Key Topics in Orthopaedic Surgery

Key Topics in Ear, Nose and Throat

KEY TOPICS IN
PAEDIATRICS

A.E.M. DAVIES
MB ChB MRCP (UK)
Senior Registrar in Child Health, Institute of Child Health, Bristol Royal Hospital for Sick Children, Bristol, UK

A.L. BILLSON
BA MBBS MRCP (UK)
Lecturer in Child Health, Queen's Medical Centre, Nottingham, UK

Consultant Editor

H.R. JENKINS
MA MD FRCP
Consultant Paediatric Gastroenterologist, University Hospital of Wales, Cardiff, UK

First published 1994

A CIP catalogue record for this book is available from the British Library.

ISBN 1 872748 58 9

BIOS Scientific Publishers Ltd
St Thomas House, Becket Street, Oxford OX1 1SJ, UK
Tel. +44 (0)865 726286. Fax. +44 (0)865 246823

DISTRIBUTORS

Australia and New Zealand
DA Information Services
648 Whitehorse Road, Mitcham
Victoria 3132

India
Viva Books Private Limited
4346/4C Ansari Road
New Delhi 110002

Singapore and South East Asia
Toppan Company (S) PTE Ltd
38 Liu Fang Road, Jurong
Singapore 2262

USA and Canada
Books International Inc.
PO Box 605, Herndon VA 22070

Typeset by Chandos, Stanton Harcourt, UK.
Printed by Information Press Ltd, Oxford, UK.

CONTENTS

ABBREVIATIONS

ACTH	adrenocorticotrophic hormone
AD	autosomal dominant
ADH	antidiuretic hormone
AFP	α-fetoprotein
AIDS	acquired immunodeficiency syndrome
ALEC	artificial lung expanding compound
ALL	acute lymphoblastic leukaemia
AML	acute myeloid leukaemia
ANA	antinuclear antibody
APLS	advanced paediatric life support
AR	autosomal recessive
AS	aortic stenosis
ASD	atrial septal defect
ASO	anti-streptolysin O
AV	atrioventricular
AVPU	alert, responsive to voice, responsive to pain, unresponsive
BPD	bronchopulmonary dysplasia
CAH	congenital adrenal hyperplasia
CAPD	continuous ambulatory peritoneal dialysis
CDH	congenital dislocation of the hip
CF	cystic fibrosis
CFTR	cystic fibrosis transmembrane conductance regulator
CGD	chronic granulomatous disease
CHD	congenital heart disease
CML	chronic myeloid leukaemia
CMV	cytomegalovirus
CNS	central nervous system
CONI	care of the next infant
CPK	creatine phosphokinase
CPP	cerebral perfusion pressure
CPR	cardiopulmonary resuscitation
CRP	C-reactive protein
CSF	cerebrospinal fluid
CT	computerized tomography
DCT	direct Coombs' test
DDAVP	1-deamino-8-D-arginine vasopressin
DI	diabetes insipidus
DIC	disseminated intravascular coagulopathy
DKA	diabetic ketoacidosis
DMSA	dimercaptosuccinic acid
ECG	electrocardiogram
EEG	electroencephalogram

ESR	erythrocyte sedimentation rate
ETT	endotracheal tube
FBC	full blood count
FDP	fibrin degradation product
FSH	follicle-stimulating hormone
Gal-1-put	galactose-1-phosphate uridyl transferase
G6PD	glucose-6-phosphate dehydrogenase deficiency
GFR	glomerular filtration rate
GH	growth hormone
GHRH	growth hormone-releasing hormone
GI	gastrointestinal
GMN	glomerulonephritis
GnRH	gonadotrophin-releasing hormone
GOR	gastro-oesophageal reflux
HIE	hypoxic–ischaemic encephalopathy
HIV	human immunodeficiency virus
HSP	Henoch–Schönlein purpura
HSV	herpes simplex virus
HUS	haemolytic uraemic syndrome
ICP	intracranial pressure
IDDM	insulin-dependent diabetes mellitus
IGF-1	insulin-like growth factor 1
IPPV	intermittent positive-pressure ventilation
IRT	immunoreactive trypsin
ITP	idiopathic thrombocytopenic purpura
ITU	intensive therapy unit
IUGR	intrauterine growth retardation
IVH	intraventricular haemorrhage
IVU	intravenous urography
JCA	juvenile chronic arthritis
JVP	jugular venous pressure
LBW	low birth weight
LGA	large for gestational age
LH	luteinizing hormone
LIP	lymphocytic interstitial pneumonitis
LP	lumbar puncture
LVH	left ventricular hypertrophy
MC&S	microscopy, culture and sensitivity
MCAD	medium-chain acyl CoA dehydrogenase
MCNS	minimal change nephrotic syndrome
MCUG	micturating cystourethrogram
MEN	metabolic endocrine neoplasia
MIBG	meta-iodobenzylguanidine
MMR	measles, mumps, rubella
MRC	Medical Research Council

MRI	magnetic resonance imaging
MSU	mid-stream urine
NAI	non-accidental injury
NEC	necrotizing enterocolitis
NHL	non-Hodgkin's lymphoma
NSAID	non-steroidal anti-inflammatory drug
OFC	occipito-frontal circumference
OPCS	Office of Population Censuses and Surveys
PA	pulmonary atresia
PCP	*Pneumocystis carinii* pneumonia
PCV	packed cell volume
PDA	patent ductus arteriosus
PET	pre-eclamptic toxaemia
PFC	persistent fetal circulation
PKU	phenylketonuria
PS	pulmonary stenosis
PT	prothrombin time
PTH	parathyroid hormone
PTT	partial thromboplastin time
PUO	pyrexia of unknown origin
PVL	periventricular leucomalacia
RAH	right atrial hypertrophy
RAST	radioallergosorbent test
RDS	respiratory distress syndrome
ROP	retinopathy of prematurity
RSV	respiratory syncytial virus
RTA	renal tubular acidosis
RVH	right ventricular hypertrophy
SBE	subacute bacterial endocarditis
SCID	severe combined immunodeficiency
SGA	small for gestational age
SIDS	sudden infant death syndrome
SLE	systemic lupus erythematosus
SPA	suprapubic aspirate
SSPE	subacute sclerosing panencephalitis
SVT	supraventricular tachycardia
T_4	thyroxine
T_3	tri-iodothyronine
TA	tricuspid atresia
TBG	thyroid-binding globulin
TGA	transposition of the great arteries
TIBC	total iron-binding capacity
TOF	tracheo-oesophageal fistula
TPN	total parenteral nutrition
TSH	thyroid-stimulating hormone

TT	thromboplastin time
U&E	urea and electrolytes
UKCCR	UK Co-ordinating Committee on Cancer Research
UKCCSG	UK Children's Cancer Study Group
US	ultrasound
UTI	urinary tract infection
VF	ventricular fibrillation
VIII RAG	factor VIII-related antigen
VLBW	very low birth weight
VSD	ventricular septal defect
VTEC	verotoxin-producing *E. coli*
VUR	vesicoureteric reflux
WBC	white blood cell count

PREFACE

This book contains 100 chapters, each providing a concise account of a topic considered to be central to a general paediatric curriculum. The text is aimed primarily at candidates approaching Part II MRCP in Paediatrics, but will also be of great value to those undertaking Part I MRCP in Paediatrics (recently introduced) and the Diploma of Child Health. Medical students and nurses may also find it a useful source of information.

It is assumed that the reader will already have acquired a basic knowledge of paediatrics. The book is therefore intended to act as a revision text, although references to more detailed texts and articles are included for many topics.

We would like to thank our many teachers, past and present, our publishers, BIOS, and the series editors for their help and encouragement.

Anne E.M. Davies
Amanda L. Billson

ABDOMINAL PAIN – ACUTE

Acute abdominal pain in childhood may be difficult to assess, especially in the young infant. The general condition of the child must be assessed, and both medical *and* surgical causes must be considered. Particular difficulties arise when more than one pathology exists, e.g. acute appendicitis with diabetic ketoacidosis. Most hospital admissions with abdominal pain are due to acute appendicitis or non-specific abdominal pain.

Common causes of abdominal pain

- Acute appendicitis.
- Non-specific abdominal pain (includes mesenteric adenitis).
- Constipation.
- Gastroenteritis.
- Urinary tract infection.
- Pneumonia.
- Acute intestinal obstruction (e.g. intussusception, volvulus).
- Diabetic ketoacidosis.

Rarer causes of abdominal pain

- Duodenal ulceration and Meckel's diverticulum.
- Henoch–Schönlein purpura.
- Testicular torsion.
- Lead or iron poisoning.
- Acute porphyria.
- Pancreatitis.
- Ovarian cysts, haematocolpos.
- Sickle cell crisis

Clinical assessment

A detailed history should be taken, enquiring about gastrointestinal and other associated symptoms, e.g. fever, cough, rash, dysuria. Examination should include assessment of the child's state of hydration and a calm and careful examination of the abdomen, including hernial orifices and external genitalia. Repeated or unnecessary rectal examinations should be avoided.

Investigation

Selected investigations may be required:

- Full blood count (FBC). A neutrophilia is often found in urinary tract infection (UTI), acute appendicitis, etc., but is non-specific. Sickle cell screen may be indicated.
- Urea and electrolytes (U&E), creatinine, glucose, amylase.
- Blood culture.
- Urine for culture, protein, glucose.
- Abdominal radiography. Erect and supine, looking for signs of intestinal obstruction, intussusception.

- Chest radiography. Lower lobe pneumonia may cause abdominal pain.
- High-resolution US in difficult cases.
- Laparoscopy in selected difficult cases.

Management

Both medical and surgical management depends on the underlying diagnosis. Intravenous fluid replacement, with intravenous colloid and antibiotics, may be indicated for the resuscitation of the patient prior to surgery, and attention should be paid to administration of appropriate analgesia.

Three common and important causes of abdominal pain are discussed in more detail below:

1. Acute appendicitis. Anorexia, vomiting and abdominal pain occur. The pain may be central or in the right iliac fossa, and is usually accompanied by localized tenderness with rebound and guarding. There may be diarrhoea if there is a pelvic appendicitis, and only a low-grade fever unless perforation has occurred. The pre-school child is particularly difficult to assess, and antibiotic administration for presumed ear and throat infections may have modified the clinical course – this group may present late, with an appendix abscess or perforation. The white cell count and abdominal radiographs will not distinguish appendicitis from other causes of abdominal pain, and repeated careful clinical assessment is required before making the decision to operate.

2. Non-specific abdominal pain (NSAP). This term has been used in recent years to include the group of children with acute abdominal pain which may be similar to acute appendicitis, but is more diffuse and variable on examination, and is unaccompanied by rebound or guarding. Repeated clinical assessment, combined with relevant investigation, must exclude more definite pathology, including acute appendicitis, UTI, intestinal obstruction, etc. Those children with mesenteric adenitis will be included in the group labelled NSAP, since this is an unreliable diagnosis to make clinically, and is often diagnosed at operation when enlarged mesenteric lymph nodes with free peritoneal fluid are found in the presence of a normal appendix. Non-specific abdominal pain usually settles over 24-48 hours.

3. Intussusception. This is caused by the invagination of one segment of bowel into another, commonly of terminal ileum into the caecum. The lead-point may be a Meckel's diverticulum, polyp, duplication cyst or an enlarged Peyer's patch. The child typically presents between 6 and 12 months with a short history of episodes of screaming, irritability and pallor, often drawing up the legs in pain. Vomiting, sometimes bile stained, may occur, and later blood and mucus may appear in the stools. The sausage-shaped mass of an ileo-colic intussusception is often palpable in the right hypochondrium. Plain radiographs indicate small bowel obstruction and occasionally the soft-tissue shadow of the intussusception. Barium enema confirms the diagnosis, and a hydrostatic reduction may be attempted by the column of barium, provided there is no peritonism and the child has been resuscitated. Successful reduction is more likely if the history has been short. Surgery is indicated if there is shock or peritonitis, or if barium reduction fails.

Further reading

Stringer MD, Drake DP. Diagnosis and management of acute abdominal pain in childhood. *Current Paediatrics,* 1991; **1**: 2-7.

Williams N. Acute appendicitis in the pre-school child. *Archives of Disease in Childhood,* 1991; **66**: 1270-2.

Related topics of interest

Abdominal pain – recurrent (p. 4)
Constipation (p. 101)
Lower respiratory tract infection (p. 211)
Urinary tract infection (p. 322)
Vomiting (p. 328)

ABDOMINAL PAIN – RECURRENT

Recurrent abdominal pain may be defined as more than three attacks of pain occurring during a period of at least 3 months. This symptom affects at least 10% of school children, and girls are more often affected than boys after puberty. There is no relationship with intelligence or social class, but siblings and parents may have had similar symptoms. The majority of children (95%) will have non-organic or functional pain, but in approximately 5% there will be clear organic pathology, and the possibility of organic illness should always be considered. Clearly, organic and non-organic factors may coexist, and both the clinical and the psychological assessments are therefore important in deciding on appropriate investigation and management.

Problems

- Exclusion of organic pathology.
- Poor school attendance.
- Associated psychosomatic disease.
- Further symptoms in adult life.

Organic causes of recurrent abdominal pain

- Urinary tract infection, urinary calculi, pelviureteric obstruction.
- Coeliac disease.
- Inflammatory bowel disease.
- Pancreatitis.
- Peptic ulceration.
- Constipation.
- Migraine.
- Gastritis and oesophagitis.
- Gall stones and cholecystitis.
- Recurrent intestinal obstruction, e.g. malrotation, volvulus.
- Ovarian cysts or tumours.
- Lead poisoning.
- Porphyria.
- Referred pain from pleura, spine, pelvis, etc.

History

A detailed history is essential, asking about the timing, frequency, nature and radiation of the pain, and its associated symptoms (non-organic pain may be associated with pallor, nausea and vomiting, and may occasionally wake the child at night). Non-organic pain may occur on school days only. In general, pain that is in a distant site from the umbilicus is suggestive of organic pathology. There may be a family history of recurrent abdominal pain, migraine, peptic ulceration or irritable bowel syndrome. A full social enquiry may reveal sources of emotional stress in the family, and written details from the child's school often provide additional information about the child's academic or social performance.

Examination

An assessment of longitudinal growth and pubertal staging should be made, followed by a thorough physical examination. The child is rarely seen during an episode, but it is helpful to arrange this if possible.

Investigation

The chosen investigations will be guided by the clinical assessment, and if non-organic symptoms are suspected a full blood count, erythrocyte sedimentation rate (ESR), urine analysis and culture may provide an initial 'screen' for disease, and more detailed tests are unnecessary. However, if the history is concerning or atypical further investigations may be needed.

- FBC, ferritin, ESR.
- U&E, creatinine.
- Liver function tests, amylase.
- Urine for glucose, protein and culture.
- Abdominal radiographs, preferably during an episode.
- US scan of abdomen and renal tract.
- Intravenous pyelogram.
- Barium contrast studies, e.g. meal and follow-through, enema.
- Oesophageal pH monitoring.
- Upper gastrointestinal endoscopy and biopsy.

Management

In non-organic pain, the first consultation should provide the reassurance that the physical examination is normal, and that a serious disease is unlikely. The gentle suggestion that the problem may be functional is sometimes appropriate at this stage. At subsequent interviews, symptoms can be reviewed with the results of investigations, leading to stronger reassurance and a fuller discussion about the basis of functional pain. Reassurance that this symptom occurs commonly in normal children, combined with suggestions to avoid reinforcing the child's response to pain in the home setting, is helpful. The child should be reviewed frequently initially – formal psychotherapeutic help may be required if symptoms are not improving.

Further reading

Apley J. *The Child with Abdominal Pains*. London: Blackwell Scientific Publications, 1975.
Murphy MS. Management of recurrent abdominal pain. *Archives of Disease in Childhood,* 1993; **69:** 409–11.

Related topics of interest

Abdominal pain – acute (p. 1)
Headache (p. 162)
Vomiting (p. 328)

ACUTE HEMIPLEGIA

An acute onset of hemiplegia is a serious event, and may arise from a wide variety of underlying causes. Detailed investigations must be undertaken to establish aetiology, although in a few cases no cause will be found.

Aetiology

1. Haemorrhage. Bleeding may arise from an arteriovenous malformation or aneurysm (congenital, traumatic or mycotic). Generalized bleeding disorders (thrombocytopenia, haemophilia, disseminated intravascular coagulation) or administration of anticoagulants may be responsible. Bleeding into a cerebral tumour may cause hemiplegia in a previously asymptomatic patient.

2. Arterial occlusion. This usually occurs as a result of occlusion of a major cerebral artery, e.g. internal carotid or middle cerebral artery. Thrombus may form as a result of infection (cervical adenitis, tonsillar infection), cyanotic congenital heart disease (polycythaemia and iron deficiency anaemia), sickle cell disease or hypertension. Progressive arterial occlusion occurs in moya-moya disease, in which recurrent episodes of hemiplegia, transient dysphasia and seizures may occur. Emboli may form from mural thrombus in cyanotic congenital heart disease with arryhthmia, acute endocarditis and, rarely, cardiac myxoma. Smaller arteries may be involved in collagen vascular disease, e.g. polyarteritis nodosa, systemic lupus erythematosus.

3. Trauma. Head injury may cause hemiplegia as a result of epidural or subdural haemorrhage. A blunt injury to the paratonsillar area may cause a thrombosis of the internal carotid artery. Stretching of the internal carotid artery against the upper cervical vertebrae may occur when there is overextension of the neck with head rotation, and both thrombi and emboli may form. Air embolus after surgical intervention and fat embolus after long bone fractures are rare events.

4. Venous occlusion. Sterile venous thrombosis may occur during dehydration, cyanotic congenital heart disease, diabetic coma and protein S or C deficiency. Infected thrombus may be responsible if there is focal intracranial infection, e.g. purulent meningitis, mastoiditis.

5. *Infection.* Acute hemiplegia may occur in bacterial meningitis as a result of thrombosis in cortical veins or arteries. The direct focal cerebral invasion of herpes simplex virus may be responsible for an acute hemiplegia, but a post-infectious encephalitis, e.g. measles or varicella, may also be involved.

6. *Migraine.* Neurological deficits may occur in complicated forms of migraine, and a hemiplegia may be accompanied by sensory changes, dysarthria or aphasia. Although spontaneous recovery is the rule, persistent weakness may rarely occur.

7. *Epilepsy and post-ictal states.* Acute hemiplegia may follow convulsions (including complicated febrile convulsions) and status epilepticus.

Clinical assessment

The history must establish the presence of any relevant chronic illness, e.g. congenital heart disease or sickle cell disease. There may be a recent history of trauma, infectious disease or vaccination. A family or past medical history of migraine or epilepsy may be important. The mode of onset of the deficit and the presence of associated symptoms is also important, e.g. the sudden onset of headache with rapidly evolving hemiplegia may suggest an intracranial haemorrhage. Physical examination may reveal signs of a generalized bleeding disorder, trauma or local or systemic infection (there may be extracardiac manifestations of infective endocarditis). Cardiological assessment, including blood pressure measurement, is mandatory. Detailed neurological examination may establish the likely vessel of occlusion, e.g. a contralateral hemiparesis with legs affected more than arms in occlusion of the anterior cerebral artery.

Investigation

- FBC, ESR, coagulation studies, haemoglobin electrophoresis. Protein C, S and antithrombin III estimation.
- Autoantibodies.
- Microbiology: blood culture, mid-stream urine (MSU), throat swab. Consider lumbar puncture (LP) only if raised intracranial pressure has been excluded. Viral studies.
- U&E, creatinine.
- Radiography of skull and cervical spine.
- EEG.
- ECG.

- Computerized tomography (CT) or magnetic resonance imaging (MRI) of head and cervical spine.
- Angiography or digital subtraction angiography to delineate vessels accurately.

Management

The initial management of acute hemiplegia depends on the underlying aetiology, aiming to arrest the disease process while limiting intracranial damage, e.g. in a patient with herpes simplex encephalitis, intravenous acyclovir is accompanied by close attention to fluid balance, management of raised intracranial pressure and treatment of seizures. If there is residual neurological impairment, physiotherapy should be instituted.

Further reading

Brett EM. Vascular disorders including migraine. In: Brett EM, ed. *Paediatric Neurology*, 2nd edn. Edinburgh: Churchill Livingstone, 1991; 545–65.

Related topics of interest

Headache (p. 162)
Seizures (p. 288)

ACUTE RENAL FAILURE

Acute renal failure occurs when a sudden decrease in renal function leads to loss of biochemical homeostasis. Oliguria or anuria leads to accumulation of nitrogenous waste products and disturbance of water and electrolyte balance. It may occur as a result of renal hypoperfusion (prerenal), parenchymal damage (renal) or obstruction of the renal tract (post-renal). If not corrected, renal hypoperfusion will lead to acute tubular necrosis and, if the insult is severe, cortical necrosis will follow. Acute on chronic renal failure may be precipitated by dehydration or an intercurrent infection.

Problems

- Fluid overload (cardiac failure, oedema, hypertension).
- Hyperkalaemia.
- Acidosis.
- Risk of infections.

Aetiology

1. Prerenal (renal hypoperfusion)

- Hypovolaemia, e.g. severe gastroenteritis, nephrotic syndrome, burns.
- Septicaemic shock.
- Cardiac failure.
- Renal vein thrombosis.

2. Renal

- Acute glomerulonephritis.
- Severe pyelonephritis.
- Cortical ischaemia, e.g. shock, haemorrhage, trauma.
- Nephrotoxins, e.g. gentamicin, gold, methotrexate.
- Haemolytic uraemic syndrome.

3. Post-renal (obstructive uropathy)

- Posterior urethral valves.
- Uric acid crystals, e.g. tumour lysis syndrome in treatment of leukaemia.
- Ureteric obstruction, e.g. stones, ureteroceles.

Clinical features

The history and examination may suggest the aetiology. Features of acute renal failure include oliguria (urine output less than 200 ml/m^2/day), oedema and fluid overload, acidotic breathing and drowsiness. Acute hypertensive encephalopathy may occur.

Investigation

- Urine. Haematuria and/or proteinuria suggest underlying renal disease. Microscopy will confirm the presence of red cells and may identify casts. Pyuria suggests infection, which should be confirmed on urine culture.

- Blood and urine U&E, creatinine and osmolality. Urea and creatinine will be raised in line with the severity of renal failure. There may be dilutional hyponatraemia, and hyperkalaemia will be exacerbated by acidosis. In prerenal failure the urinary sodium concentration will be low (< 30 mmol/l), whereas in renal failure the kidneys tend to leak sodium and the concentration will be higher (> 30 mmol/l). In prerenal failure the urine is appropriately concentrated (urine to plasma osmolality ratio > 1.15, urine to plasma urea ratio > 10), but in renal failure the ratios will be lower.
- FBC and clotting studies. The haematocrit may be a useful guide to hypovolaemia. There will be thrombocytopenia and deranged clotting if there is associated disseminated intravascular coagulopathy (DIC).
- Acid–base status. Metabolic acidosis.
- Liver function tests, calcium, phosphate, albumin.
- Blood culture.
- Radiology. Renal tract US may detect dilatation of the renal tract in obstructive uropathy. Renal size and cortical thickness may reveal intrinsic renal disease (dysplasia, polycystic disease, acute on chronic renal failure). Renal isotope scans are useful to detect underlying chronic renal disease and to follow recovery from the acute episode. Cystoscopy and micturating cytourethrography are indicated for the diagnosis and treatment of posterior urethral valves. Chest radiographs may show pulmonary oedema and cardiomegaly if fluid overload precipitates heart failure.

Management

1. Fluids. Hypovolaemia should be corrected immediately with plasma 10–20 ml/kg. If the patient is hypoalbuminaemic (e.g. nephrotic syndrome) use 20% albumin. Central venous pressure monitoring enables more accurate assessment of intravascular volume. When fluid replacement is adequate and oliguria persists, diuretics such as frusemide may improve urine output. Accurate fluid balance is essential and weight should be monitored to avoid fluid overload. Urinary catheterization allows accurate measurement of urine output. Maintenance fluids should replace insensible losses (300 ml/m^2/day) plus urine output.

2. *Hyperkalaemia.* This can lead to arrhythmias and cardiac arrest and so should be carefully monitored. ECG changes (peaked T waves, depressed R waves, prolonged QRS and PR intervals) or a serum potassium > 7 mmol/l are indications for emergency treatment while dialysis is arranged:

- Intravenous calcium gluconate 0.5 ml/kg over 2–4 minutes (reduces the cardiotoxic effect of hyperkalaemia) followed by dextrose and insulin, which reduces the serum potassium by promoting uptake into cells. The blood glucose level should be monitored carefully to avoid hypo/hyperglycaemia.
- Nebulized salbutamol given 2-hourly will reduce the serum potassium and can be used in conjunction with dextrose and insulin.
- Calcium resonium orally or rectally increases the elimination of potassium from the body.

3. *Acidosis* will correct with the reduction of uraemia. If severe, correct with sodium bicarbonate.

4. *Hypertension* occurs as a result of salt and fluid overload. If strict fluid balance and diuretics are inadequate antihypertensive drugs such as beta-blockers or vasodilators can be used. If hypertension remains severe and uncontrolled with drug treatment, dialysis is indicated.

5. *Risk of infections.* Treat suspected infections promptly with broad-spectrum antibiotics to prevent catabolism. Use aminoglycosides with caution.

6. *Nutrition.* Intake should be adequate to minimize catabolism. Uraemia causes anorexia and vomiting so intravenous feeding may be necessary.

7. *Dialysis.* Peritoneal dialysis is the method of choice in children. Indications for dialysis include uncontrolled hyperkalaemia, acidosis or fluid overload.

Outcome

The outcome depends on the underlying cause. A child with acute tubular necrosis secondary to hypovolaemia will recover renal function in 7–14 days. Polyuria occurs during the recovery phase and careful monitoring of fluid balance and electrolytes is necessary. Acute renal failure secondary to rapidly progressive glomerulonephritis or renal cortical necrosis has a worse prognosis.

Further reading

Robson AM, Cameron JS. Acute renal failure in the neonate and older child. In: Cameron S, Davison AM, Grunfeld JP, Kerr D, Ritz E, eds. *Oxford Textbook of Clinical Nephrology*. Oxford: Oxford University Press, 1992: 1110–23.

Related topics of interest

Chronic renal failure (p. 75)
Hypertension (p. 177)
Shock (p. 291)

ADRENAL DISORDERS

The adrenal cortex produces glucocorticoid, mineralocorticoid and sex hormones. Cortisol, the principal glucocorticoid, modulates stress and inflammatory responses. It is a potent stimulator of gluconeogenesis and antagonizes insulin. Aldosterone, the principal mineralocorticoid, causes increased sodium reabsorption and potassium and hydrogen ion loss from the distal renal tubule. Androstenedione is converted by the liver to testosterone in boys and oestrogen in girls, and production increases markedly at puberty. Cortisol and androgen production are under diurnal control by the hypothalamo-pituitary axis (adrenocorticotrophic hormone–ACTH). Aldosterone is released in response to angiotensin II, produced following renal renin release and subsequent pulmonary angiotensin I conversion. Clinical disease results from relative excess or lack of hormones.

Congenital adrenal hyperplasia (CAH)

This term encompasses a group of autosomal recessive disorders of adrenal corticosteroid biosynthesis, resulting from deficiency of one of five enzymes in the cholesterol to cortisol pathway. By far the most common is 21-hydroxylase deficiency, with a prevalence of 1 in 5000 live births. This enzyme is also involved in the cholesterol to aldosterone pathway, and a deficiency will lead to aldosterone deficiency and a salt-losing crisis.

Presentation

- Ambiguous genitalia at birth (virilized female).
- Vomiting, dehydration and salt loss (salt-losing crisis) in the first few weeks of life.
- Collapse during stress, e.g. intercurrent illness.
- Virilization in early childhood. Males may present with false precocious puberty. Females may develop polycystic ovary syndrome (acne, hirsutism, delayed menarche, irregular periods).
- Hypertension.

Investigation

- *Cortisol low or normal.* Unstressed levels may be normal.
- *ACTH raised.* Reduced feedback inhibition of ACTH by cortisol. ACTH stimulates abnormal production of other steroids (e.g. androstenedione), leading to virilization.
- *Raised plasma steroid precursors.* Accumulation proximal to the block (e.g. plasma 17-OH-progesterone).
- *Urinary steroid metabolites,* e.g. pregnanetriol is main metabolite of 17-OH-progesterone.
- Aldosterone low, renin raised if salt losing.

Management

1. Emergency treatment of a salt-losing crisis. All babies with ambiguous genitalia should be considered to be at risk of a salt-losing crisis. The first sign may be a rising potassium. Mineralocorticoid and salt replacement before the serum sodium falls will avert a hypotensive crisis. CAH, particularly in boys, may present with a salt-losing crisis. Dehydration and hypotension require plasma volume expansion with dextrose and saline. Boluses of i.v. aldosterone may be needed. Hydrocortisone should be given intravenously.

2. *Hormone therapy.* Hydrocortisone is given to replace cortisol secretion and to suppress excessive ACTH production. The dose must be increased during periods of stress, e.g. intercurrent infection, general anaesthesia. Growth and pubertal development should be monitored closely. If salt-losing, then mineralocorticoid replacement (oral fludrocortisone) with oral sodium supplements is essential. Renin levels should be monitored to ensure adequate mineralocorticoid replacement.

3. *Ambiguous genitalia.* Sex assignment with chromosomes and pelvic ultrasound is extremely important. Parents are often devastated by the problem but should be encouraged not to register the birth or name the child (particularly with an ambiguous name, e.g. Leslie) until the sex has been decided. Surgery (clitoral reduction, labial separation) may be necessary to achieve functionally and cosmetically normal external genitalia. Other causes of ambiguous genitalia are rare, e.g. exogenous androgens, maternal or fetal androgen-secreting tumours, testosterone synthesis defects, true hermaphroditism. The karyotype is not always the most important factor in sex assignment in these conditions, the decision being based on the possibility of achieving the most normal external genitalia, and reproductive capacity where possible, with surgery and hormonal therapy.

Cushing's syndrome

This is uncommon in children. Clinical features are due to excessive glucocorticoid, leading to protein catabolism, increased carbohydrate production, fat accumulation and potassium loss (moon face, thin skin, easy bruising, hypertension (60%), hirsutism, obesity, buffalo hump, muscle weakness, diabetes, osteoporosis, aseptic necrosis of the hip, pancreatitis). The commonest cause is corticosteroid treatment. Rare causes of endogenous glucocorticoid excess include adrenal hyperplasia, adrenal cortex tumours and pituitary ACTH hypersecretion (Cushing's disease = ACTH-secreting pituitary tumour). Management depends on removing the cause either medically or surgically.

Adrenocortical failure

Failure of the adrenal cortex may be due to a primary adrenal disorder (Addison's disease (autoimmune), CAH, adrenal haemorrhage), or it may be secondary to hypothalamo-pituitary disorders leading to ACTH insufficiency (e.g. tumours of the hypothalamo-pituitary axis, congenital hypopituitarism, surgery or radiotherapy and corticosteroid treatment). There may be complete failure of the adrenal cortex or selective failure of glucocorticoid, mineralocorticoid or androgen synthesis. Onset may be insidious with fatigue, weight loss, nausea and hyperpigmentation, or acute with an Addisonian crisis (hypotension, hypoglycaemia, hyperkalaemia) during a period of stress such as an infection. Management of acute adrenocortical insufficiency consists of resuscitation with plasma, hormone replacement and correction of hypoglycaemia and electrolyte disturbances.

Primary hyperaldosteronism (Conn's syndrome)

This is very rare in children and is usually due to a zona glomerulosa adenoma. Features include sodium and water retention, hypertension, hypokalaemia, muscle weakness, polyuria and impaired growth. There is a hypochloraemic, hypokalaemic alkalosis which is also found in Bartter's syndrome. In this latter condition hyperaldosteronism is secondary to juxtaglomerular apparatus hypertrophy with raised renin levels and, in contrast to Conn's syndrome, the blood pressure is normal.

Further reading

Appan S, Hindmarsh, PC, Brook CGD. Monitoring treatment in congenital adrenal hyperplasia. *Archives of Disease in Childhood,* 1988; **64:** 1235–9.

Brook CGD. The adrenal gland. In: Brook CGD, *A Guide to the Practice of Paediatric Endocrinology.* Cambridge University Press, 1993: 97–118.

Brook CGD. Intersex. In: Brook CGD, *A Guide to the Practice of Paediatric Endocrinology.* Cambridge University Press, 1993: 1–17.

Pagon RA. Diagnostic approach to the newborn with ambiguous genitalia. *Pediatric Clinics of North America*, 1987; **34(4):** 1019–31.

Related topics of interest

Shock (p. 291)
Vomiting (p. 328)

ALLERGY

Allergy may be defined as a state of altered reactivity to a particular substance which is mediated by an immunological response to that specific allergen or antigen. The reaction is usually reproducible. The substance causing the reaction may have been ingested, inhaled or come into physical contact with skin or mucous membranes.

Problems

- Asthma.
- Eczema.
- Allergic rhinitis.
- Food intolerance (causing gastrointestinal problems such as diarrhoea and vomiting).
- Drug allergy.
- Anaphylaxis.

There are four categories of allergic reaction:

Types of allergic reaction

1. Type 1: immediate hypersensitivity (anaphylactic) reaction. Allergens react with reaginic (IgE) antibodies on the surface of mast cells or basophils, causing immediate release of vasoactive substances. Reactions may occur within minutes, e.g. allergic rhinitis, urticaria or food allergy, and in its most severe form, acute anaphylaxis, there is a profound reaction with shock, hypotension and airway oedema.

2. Type 2: cytotoxic reaction. Circulating antibody reacts with antigen which is bound to a cell surface, and complement is often involved in the subsequent cell damage. Such reactions include rhesus and ABO isoimmunization, autoimmune haemolytic anaemia and post-streptococcal glomerulonephritis.

3. Type 3: immune complex mediated (Arthus phenomenon). Soluble immune complexes are formed from combined antibody-activated complement and antigen which is present in excess. This reaction occurs within hours of exposure. The excessive immune complexes persist, attracting neutrophils to the affected tissue and causing more tissue damage, e.g. vasculitis in systemic lupus erythematosus, extrinsic allergic alveolitis.

4. Type 4: delayed or cell-mediated hypersensitivity. These reactions take at least 24 hours to develop, are mediated by sensitized T lymphocytes and do not involve antibody or complement. Lymphokines are released which cause local inflammation and infiltration by lymphocytes and macrophages. Examples are organ transplantation rejection and tuberculin hypersensitivity.

Allergic rhinitis

Intense irritation in the nose with congested conjunctivae, sneezing and a clear nasal discharge are the typical features. A post-nasal drip stimulates the cough reflex. Sensitivity may be to grass and tree pollen. There may be swelling and discoloration below the eyelids (allergic shiners) and excoriation of the nostrils and upper lip. Antihistamines and oral adrenergic agents may be combined with topical therapy such as sodium cromoglycate or corticosteroid preparations.

Food intolerance

This is a reproducible response to a specific food which is not psychologically based. Unless its immunological basis can be proven (rare), it is inappropriate to label it as a food allergy in the true sense. Cow's milk protein intolerance is one of the most common food intolerances in childhood. Symptoms such as loose stools, vomiting, rash and poor weight gain occur after the introduction of cow's milk. The child may have various atopic conditions, or there may be a family history. Treatment is the total exclusion of dietary cow's milk protein, using a soya-based or casein hydrolysate milk formula. Such diets should be fully supervised, since other milk substitutes may lack calcium and vitamins. Most children tolerate cow's milk by 3 years of age.

Anaphylaxis

Initial clinical features are irritation and itching of the mouth, malaise, weakness and vomiting, followed by bronchospasm, upper airway oedema, hypotension and shock. Emergency treatment is to secure the airway, with cricothyrotomy or tracheostomy if necessary, ensure oxygenation and to administer intramuscular or subcutaneous adrenaline at a dose of 0.1ml/kg of 1:10 000 solution. Repeated adrenaline doses can be used if the effect wears off. Intravenous fluids are needed to support the circulation. Steroids, antihistamines and nebulized salbutamol may also be given.

Investigation of allergic conditions

1. Skin tests. The skin weal and flare response to various extracts of allergens is tested. False-positive and false-negative results are common, there is variability of the skin reaction depending on site and potency of allergen, and non-IgE-mediated reactions are not detected.

2. *RAST and IgE.* The radioallergosorbent (RAST) test is a semiquantitative assay for allergen-specific IgE. It may be used when skin testing is impossible, e.g. widespread eczema. There are similar problems with interpretation as for skin tests. Total IgE level is related to duration and exposure to the allergen, and occasionally helps in the distinction of atopic and non-atopic disease.

3. *Provocation tests.* These may apply the allergen directly on to the nasal mucosa or into the bronchi by inhalation, or may involve ingestion of a suspected food. Responses may be clinical symptoms or changes in measured lung function.

Further reading

David TJ. *Food and Food Additive Intolerance in Childhood.* Blackwell Scientific Publications, Oxford, 1993.

Trigg CJ, Davies RJ. Allergic rhinitis. *Archives of Disease in Childhood,* 1991; **66:** 565–7.

Related topics of interest

Asthma (p. 29)
Cardiopulmonary resuscitation (p. 52)
Eczema (p. 127)

ANAEMIA

The haemoglobin level is high at birth (15–19 g/dl) and falls to a nadir at about 2–3 months of age, but it seldom falls to below 10 g/dl in healthy infants. Anaemia may result from failure of red cell production, increased red cell breakdown or haemorrhage. It is often asymptomatic, but clinical features include tiredness, lethargy and pallor. Breathlessness and cardiac failure are more common if the onset of anaemia is acute (e.g. acute haemolysis). The commonest cause of anaemia in childhood is iron deficiency.

Aetiology

1. Failure of production

- Aplastic anaemia (congenital, acquired, e.g. drugs, crises).
- Iron deficiency (dietary lack, chronic blood loss).
- B_{12} deficiency (Crohn's disease).
- Folate deficiency (coeliac disease, anticonvulsants, haemolysis).
- Bone marrow invasion (leukaemia, tuberculosis).
- Erythropoietin deficiency (renal disease).
- Chronic infection.

2. Increased breakdown (haemolysis)

(a) Hereditary

- Membrane defects (spherocytosis).
- Haemoglobinopathies (sickle cell anaemia, thalassaemia).
- Enzyme deficiencies (glucose-6-phosphate dehydrogenase (G6PD) deficiency, pyruvate kinase deficiency).

(b) Acquired

- Immune (rhesus, ABO, blood transfusion incompatibility).
- Autoimmune (drugs, e.g. methyldopa; infections, e.g. mycoplasma).

(c) Other

- Burns.
- Artificial heart valves.
- Haemolytic uraemic syndrome.
- Snake venom.

3. Blood loss

- Perinatal (fetomaternal, twin to twin, placental/cord accidents, haemorrhagic disease of the newborn).
- Epistaxis.
- Trauma.
- Gastrointestinal (acute/chronic).
- Haematuria.
- Bleeding disorders (haemophilia, thrombocytopenia).

Iron deficiency anaemia

Seventy-five per cent of the total body iron at birth is found in the circulating haemoglobin, with the remainder stored as ferritin and haemosiderin. This is sufficient to meet requirements for the first 4–6 months of life, and infants should be weaned before this time. Most cases of iron deficiency anaemia are due to poor dietary intake and present in late infancy and early childhood. Prolonged breast-feeding without the introduction of adequate solids will lead to iron deficiency. Infant formula milks are fortified with iron, but doorstep cow's milk is low in available iron and also causes microscopic gastrointestinal bleeding under the age of 12 months. Also children who drink a lot of milk tend to have a poor intake of solids and are frequently iron deficient. Iron absorption occurs in the duodenum and jejunum and is enhanced by gastric acid, protein and ascorbates. Only 5–10% of the dietary iron is normally absorbed, and a high intake of phytates (e.g. chapattis) or phosphates in the diet will reduce absorption further. Premature and low birth weight infants have reduced iron stores and should routinely receive supplements in the first year of life. Less common causes of iron deficiency are chronic blood loss and malabsorption. Oral iron supplements should be given for at least 3 months to correct anaemia and replace iron stores. Blood transfusion is rarely necessary.

Investigation

- *FBC.* Microcytic, hypochromic anaemia. It is important to exclude β-thalassaemia trait (microcytosis and hypochromia very marked for degree of anaemia) and anaemia of chronic disease.
- *Stools* for occult blood, faecal fat, ova, cysts and parasites, e.g. hookworms.
- *Serum iron low, total iron-binding capacity (TIBC) high, ferritin low.* Iron and TIBC are both normal in thalassaemia trait and low in chronic disease.
- *Free erythrocyte protoporphyrin* increased.

Folate and B_{12} deficiency

Megaloblastic anaemia is uncommon but is usually due to folate deficiency. Poor dietary intake may be exacerbated by rapid growth, fever, infection, diarrhoea and haemolysis, all of which increase folate requirements. Folate absorption occurs in the small intestine and is impaired in malabsorption states, e.g. coeliac disease, Crohn's disease, blind loop syndrome. Several drugs are associated with folate deficiency, e.g. phenytoin, methotrexate, cotrimoxazole.

B_{12} is combined with intrinsic factor (IF) from gastric parietal cells and is absorbed in the terminal ileum. Deficiency is uncommon as stores are sufficient for 2–3 years, but may result from inadequate intake (vegans) or failure of absorption (IF deficiency, blind loop syndrome, Crohn's disease). IF deficiency (pernicious anaemia) is rare in childhood.

Investigation

- *FBC.* Macrocytosis.
- *Red cell folate.*
- *Serum B_{12} and folate.*
- *Deoxyuridine suppression test* for folate deficiency.
- *Schilling test* for IF deficiency.

Haemolytic anaemia

The bone marrow can compensate for excessive red cell destruction by increasing erythropoiesis six- to eightfold before anaemia develops. The cause of haemolysis may be hereditary or acquired (see above). The commonest hereditary haemolytic anaemia in the UK is hereditary spherocytosis. This is an autosomal dominant membrane defect which can present at any age. Anaemia is variable, splenomegaly is common and jaundice is usually mild. Exacerbations of haemolysis are associated with intercurrent infections. Pigment gall stones are common. Aplastic crises may occur, usually precipitated by parvovirus infection or folate deficiency. Splenectomy extends the life of the red cells and may be necessary if the haemolysis is severe, but should be delayed for as long as possible.

G6PD is a red cell enzyme which protects the cell membrane from oxidant stress. G6PD deficiency is due to X-linked mutations, which are most common in African Negroes and Mediterranean and Oriental races. It may present with neonatal jaundice or with acute haemolysis on exposure to certain drugs (e.g. antimalarials, salicylates, sulphonamides), following ingestion of fava beans or during intercurrent infections.

Investigation

- *FBC, film and reticulocyte count.* Raised reticulocytes, fragmented red cells, spherocytes.
- *Serum bilirubin.* Raised unconjugated bilirubin, raised urinary urobilinogen.
- *Blood group and direct Coombs' test (DCT).* Blood group incompatibility produces haemolysis in the first days of life (see Neonatal jaundice, p. 231).
- *Osmotic fragility test.* For membrane defects.
- *Red cell enzyme assays.*
- *Haemoglobinuria.* Occurs when intravascular haemolysis leads to saturation of the haemoglobin carrier protein, haptoglobin.
- *Cold agglutinins.* Following mycoplasma infection.

Further reading

Stevens D. Epidemiology of hypochromic anaemia in young children. *Archives of Disease in Childhood,* 1991; **66**: 886–9.

Related topics of interest

Haemophilia and Christmas disease (p. 159)
Purpura and bruising (p. 278)
Neonatal jaundice (p. 231)
Sickle cell anaemia and thalassaemia syndromes (p. 295)

ANTENATAL DIAGNOSTICS

Routine screening tests are carried out in most pregnancies in the UK, and further tests are available for those at higher risk of a fetal abnormality. Recent advances in antenatal imaging and DNA techniques have widened the range of conditions which can be diagnosed antenatally. Prenatal diagnosis occasionally allows *in utero* treatment (e.g. rhesus disease), and the detection of congenital malformations which are amenable to surgery (e.g. exomphalos) allows delivery of the baby in a centre with facilities for intensive care and neonatal surgery. In severe disorders for which no treatment is available, termination of pregnancy may be indicated, so the parents must be counselled carefully and find this acceptable before extensive investigations are instituted.

Antenatal screening tests

Methods

1. Maternal FBC, blood grouping and rubella titre are carried out at the first antenatal visit. If the mother is rhesus negative the rhesus antibody titre should be monitored. Haemoglobin electrophoresis is indicated in those at high risk of haemoglobinopathy and should be routine in African Negroes (15% carry the sickle cell gene).

2. Routine ultrasound (US). Most centres perform a US scan at 16–18 weeks' gestation to confirm dates, locate the placenta and detect structural abnormalities (e.g. neural tube defects, severe skeletal dysplasia, diaphragmatic hernia, renal tract abnormalities). Polyhydramnios may indicate an underlying gastrointestinal atresia, and oligohydramnios is associated with renal abnormalities. More detailed high-resolution anomaly US scans by skilled operators can detect other abnormalities, e.g. various types of congenital heart disease and cleft lip and palate.

3. Maternal serum α-fetoprotein (AFP) is used as a screening test at 16–18 weeks for neural tube defects. Over 80% of open neural tube defects and anencephalies can be detected by a raised AFP, but other causes of a raised level include multiple pregnancy, abdominal wall defects and Turner's syndrome, so further detailed ultrasonography and amniocentesis are indicated. A low serum AFP is associated with an increased risk of Down's syndrome.

Indications for further investigation

Prenatal investigations are indicated if there is a reliable test available, the disorder is severe and there is an increased risk of fetal abnormality. Factors associated with an increased risk include:

- Maternal age over 35 years. Most centres offer amniocentesis to this group because of the increased risk of Down's syndrome. The triple test may further identify mothers at increased risk of Down's syndrome (low AFP, low serum unconjugated oestriol and high human chorionic gonadotrophin).
- Previous affected child, e.g. the recurrence risk is 1 in 4 for an autosomal recessive condition.
- Family history of an inherited condition. Investigation of a family with a single gene defect should ideally be carried out before pregnancy so that detection of biochemical carriers and DNA analysis for genetic markers can be used to estimate the risk of an affected fetus and allow genetic counselling.
- Raised maternal serum AFP.

Amniocentesis

Amniocentesis is performed at 16–18 weeks. Amniotic fluid is obtained by passing a needle through the abdominal wall and uterus under US control. Protein levels or metabolites can be measured in the amniotic fluid, e.g. raised AFP and acetylcholinesterase in neural tube defects, bilirubin in rhesus disease. This fluid also contains cells that have been shed from the fetal skin, which can be cultured and examined for chromosomal abnormalities, enzyme defects and specific gene defects. Prediction of fetal sex allows abortion of male fetuses in X-linked conditions. In some centres blood samples can be taken directly from the umbilical cord (cordocentesis) for the diagnosis of inborn errors of metabolism, haemoglobinopathies and viral infections. Cordocentesis is also used for blood transfusion in severe rhesus disease.

Problems

- Risk of miscarriage. This is about 1% following amniocentesis and about 2% following cordocentesis, compared with a spontaneous abortion rate at 16–18 weeks of 0.5%.
- Time for results. Culture of amniotic cells and analysis takes 2–3 weeks and this can be an extremely stressful time for couples. If termination of pregnancy is indicated this is relatively late in gestation.
- Small risk of haemorrhage from placenta, cord or uterine vessels, particularly following cordocentesis.
- Small risk of amnionitis.
- Risk of inducing rhesus isoimmunization. Rhesus-negative mothers should be given anti-D.
- Small increase in incidence of postural deformities, e.g. talipes, owing to amniotic fluid leakage.

Chorionic villus sampling (CVS)

CVS was initially very popular as it can be performed earlier in pregnancy than amniocentesis (8–12 weeks) and results are available within 24 hours, but the miscarriage rate is about 4% and recent publications have suggested an increased risk of limb and facial deformities following CVS so it is now only used in very high-risk pregnancies.

Fetoscopy

Direct visualization of the fetus with a fibreoptic endoscope to identify dysmorphic features or take fetal tissue samples (e.g. skin for diagnosis of epidermolysis bullosa) is now rarely used because of the improved imaging techniques. The risks of miscarriage, amniotic leakage and amnionitis are all high.

Further reading

Chitty L, Campbell S. Ultrasound screening for fetal abnormalities. In: Brock DJH, Rodeck CH, Ferguson-Smith MA, eds. *Prenatal Diagnosis and Screening*. Edinburgh: Churchill-Livingstone, 1992; 595–609.

Related topics of interest

Chromosomal abnormalities (p. 68)
Inborn errors of metabolism (p. 191)
Single gene defects (p. 299)

ARRHYTHMIAS

Sinus tachycardia

The usual heart rates per minute are:

Newborn	100–150
Toddler	85–125
3–5 years	75–115
over 5 years	60–100

A sinus tachycardia is a sinus rhythm that is faster than normal and is the normal physiological response to a requirement for increased cardiac output, e.g. exertion. The commonest cause is fever, and others include anaemia and hypovolaemia.

Sinus bradycardia

A persistently slow heart rate is uncommon in infants, but it may fall transiently, particularly during sleep, to as low as 60 beats/min. Important pathological causes include raised intracranial pressure and hypothermia.

Supraventricular tachycardia (SVT)

This is the most common abnormal tachycardia of childhood. Thirty to forty per cent of cases present within the first few weeks of life, and underlying structural heart disease is uncommon. It may present *in utero* as hydrops fetalis, or with cardiogenic shock in the neonatal period. More commonly it presents with increasing tachypnoea, poor feeding and pallor in early infancy. The older child usually presents with pallor and palpitations. The tachycardia may result from enhanced automaticity of the atrium (atrial tachycardia) or be sustained by a re-entry circuit involving an accessory atrioventricular (AV) connection (junctional tachycardia). The QRS complexes are narrow in contrast to those in ventricular tachycardias.

Junctional tachycardias

The commonest cause of SVT in childhood is an accessory connection between the ventricle and the atrium leading to an AV re-entry tachycardia. Rates of between 220 and 300 beats/min are seen during episodes of SVT. The QRS complexes are regular and narrow, with a one-to-one relationship with P waves. The accessory connection may occur anywhere around the AV ring and, if it is capable of supporting both antegrade and retrograde conduction between the atrium and ventricle, the ECG during sinus rhythm will show a short PR interval and a delta wave (Wolff–Parkinson–White syndrome). If conduction is restricted to the retrograde direction, this ventricular pre-excitation will not be seen. Occasionally there is an underlying structural abnormality, most commonly Ebstein's anomaly of the tricuspid valve.

Atrial tachycardias

Atrial tachycardia, flutter and fibrillation are uncommon in the first year of life. Predisposing factors include cardiac surgery, structural lesions in which the atria are distorted or distended and myocarditis. There is variable conduction to the ventricles resulting in irregular QRS complexes with loss of the one-to-one relationship with P waves.

Management of SVT

1. Document the tachycardia with a 12-lead ECG. Exclude ventricular tachycardia (widened QRS complexes, abnormal QRS axis for age, AV dissociation).

2. Vagal manoeuvres. Elicit the diving reflex (bag of ice on face, immerse face in cold water) or use carotid sinus massage. If unsuccessful, proceed to drug treatment.

3. Adenosine. This acts by slowing conduction through the AV node, thus disrupting the re-entry circuit. It has a rapid onset of action and a very short half-life so side-effects are transient (flushing, tachypnoea, bradycardia, complete AV block). The tachycardia may reinitiate and the dose should be increased at 2-minute intervals until a sustained response is achieved.

4. Other drugs. Verapamil is effective but it is negatively inotropic and suppresses both sinus and AV node function, leading to a risk of profound bradycardia and hypotension. It is no longer recommended in infancy and is contraindicated in the presence of beta-blockers. Digoxin has been used to treat SVT for many years. It is usually effective and is a positive inotrope but its onset of action is slow (up to 12 hours), and if given inappropriately in ventricular tachycardia it can precipitate ventricular fibrillation. If the child is not compromised by the SVT and the diagnosis is certain, intravenous digitalization is indicated if adenosine fails, but if the child is compromised or if the diagnosis is uncertain proceed to DC shock or pacing. Digoxin should not be used in Wolff–Parkinson–White syndrome. Flecainide has been recommended in the past but is not currently licensed for children

5. DC cardioversion (1–2 J/kg). Atrial arrhythmias seldom respond to vagal manoeuvres, and adenosine and digoxin merely slow the ventricular rate. DC cardioversion may be effective in restoring sinus rhythm.

6. *Oesophageal pacing* is effective but usually only available in cardiac centres.

Prevention of further episodes

Recurrent attacks may be prevented by treatment with digoxin or propranolol. Surgical ablation of the accessory connection is sometimes possible.

Heart block

Impaired conduction within the AV node and bundle of His may take three forms.

First-degree heart block

The PR interval is prolonged for age and heart rate (usually > 0.20 seconds). It occurs in up to 10% of normal children, but may be secondary to rheumatic heart disease, congenital heart disease (e.g. Ebstein's anomaly) or digitalis toxicity.

Second-degree heart block (Wenckebach phenomenon)

There is progressive prolongation of the PR interval until one P wave is not succeeded by a QRS complex.

Third-degree (complete) heart block

Complete AV dissociation may be congenital, or acquired following surgery or myocarditis. It is associated with maternal systemic lupus erythematosus and appears to be due to the transfer of maternal anti-Ro or anti-DNA antibodies.

Further reading

Cardiac emergencies. In: Advanced Life Support Group. *Advanced Paediatric Life Support.* London: BMJ Publications, 1993; 87–94.

Till JA, Shinebourne EA. Supraventricular tachycardia: diagnosis and current acute management. *Archives of Disease in Childhood,* 1991; **66:** 647–52.

Related topics of interest

Cardiopulmonary resuscitation (p. 52)
Congenital heart disease – acyanotic (p. 91)
Congenital heart disease – cyanotic (p. 95)
Heart failure (p. 168)

ASTHMA

Asthma is a chronic disease characterized by reversible airway obstruction over periods of time in response to various stimuli. It is a major cause of paediatric morbidity, affecting at least 1 in 10 children in UK, and prevalence is increasing. Hospital admissions for asthma have increased greatly during the last two decades.

Problems

- Episodic wheezing.
- Acute severe asthma.
- Poor growth.
- Psychosocial problems.

History

Diagnosis is usually based on a history of recurrent wheeze, cough, often at night, and dyspnoea. There may be clear precipitating factors such as upper respiratory tract infections, cigarette smoke, pollen, animal danders, exercise, excitement or changes in temperature. There may have been significant respiratory illness in the neonatal period, or viral bronchiolitis during the first year of life. The child may also suffer from eczema, hay fever or urticaria, or there may be atopic symptoms in family members. Previous treatments, their method of delivery and efficacy should be documented. Frequency of symptoms and the degree of interference with normal activity should be established. Systemic enquiry is important to exclude symptoms suggestive of other causes of wheeze, e.g. cystic fibrosis, immunodeficiency.

Examination

During an acute episode, tachypnoea, tachycardia, intercostal and subcostal recession with use of accessory muscles of respiration are frequent findings. Auscultation reveals widespread wheezing. However, in acute severe episodes, cyanosis with very poor air entry (silent chest) and inability to talk is a sign of life-threatening asthma. Examination between episodes is often normal, but growth should be documented, and a check made for signs of undertreated asthma (Harrison's sulcus, pectus carinatum), eczema, urticaria or signs of an alternative diagnosis, e.g. clubbing in cystic fibrosis.

Investigation

In most patients with a clear history no investigation is needed.

- Chest radiographs are needed in acute severe asthma (possible pneumothorax, lobar collapse), but not episodic wheeze if milder. Chest radiography at diagnosis in young children is wise to exclude foreign body aspiration, compression by large vessels causing wheeze, etc.

- Peak flow. Expected mean values are related to height. Bronchodilators may demonstrate reversible airway narrowing, and a home diary card is of value to monitor progress.
- Skin prick tests may help to identify allergic triggers.
- Pulse oximetry is valuable as a non-invasive assessment of oxygenation.
- Arterial blood gases. Needed in severe episodes, usually if considering mechanical ventilation.

Management

In an out-patient setting, attention should be given to avoidance of known trigger factors where possible. Passive smoking should be avoided. Energetic washing and vacuuming of carpets and soft furnishings will reduce the exposure to house-dust mite, often to good effect. Drug therapy and its method of delivery should be carefully selected according to the child's symptoms and age, and inhaler technique should always be checked. Therapy may be for prophylaxis or treatment. The main groups are:

1. β_2-Agonists: oral or inhaled terbutaline or salbutamol. Inhaled administration may be via a nebulizer, via a metered dose inhaler (MDI) or as a dry powder, e.g. Turbohaler or Diskhaler. Metered dose inhalers need good coordination and are only suitable for children older than 10. However, combined with a spacer device (Nebuhaler or Volumatic), drug delivery improves, and spacers may be used in the 4–10 year age group and in younger children when a face mask is added.

2. Sodium cromoglycate (Intal). Administration may be via a nebulizer, MDI or dry-powder inhaler.

3. Inhaled steroids. Beclomethasone dipropionate (Becotide) and budesonide (Pulmicort) may be given by MDI, nebulizer or as a dry powder; Fluticasone (Flixotide) has recently been introduced.

4. Oral steroids – prednisolone (1–2 mg/kg/day for 1–5 days).

5. Theophylline preparations. Usually a slow-release preparation is chosen, e.g. Slophyllin. Side-effects include vomiting and sleep disturbance, and blood levels are required.

Children with infrequent episodes need no regular treatment except effective β_2-agonists when wheezy. If such treatment is needed more than three times per week, or after a more severe attack, prophylaxis with sodium cromoglycate should be tried for 6 weeks. If symptoms remain uncontrolled on sodium cromoglycate, low-dose inhaled steroids can be substituted or added (usually 400 μg per day). For a few children, doses of inhaled steroid > 600 μg/day still fail to achieve control, and long-acting, β_2-agonists, e.g. salmeterol or slow-release theophylline, are required. Inhaled ipratropium bromide may also help some children. Rarely, higher inhaled steroid doses or oral prednisolone is needed to achieve control. Growth should be carefully monitored.

Management of acute severe asthma (status asthmaticus)

For many acute asthmatic episodes in children, nebulized β_2-agonists alone or with a short 3–5 day course of prednisolone brings effective relief. For severe episodes, nebulized bronchodilators should be given and repeated every hour or continuously if necessary. Nebulizers should always be oxygen driven. Oral or intravenous steroids should be given (prednisolone 1–2 mg/kg or hydrocortisone 4 mg/kg, repeated 6-hourly). If response is poor, intravenous aminophylline given as a loading dose followed by a continuous infusion (6 mg/kg load over 20 minutes then 1 mg/kg/h) may be added, omitting the loading dose if theophylline has been given during the previous 12 hours. Intravenous β_2-agonists may be added if there is a poor clinical response, monitoring the condition on an intensive therapy unit (ITU), where treatment should be maximized, and consideration given to mechanical ventilation.

Further reading

Warner JO. Asthma: a follow up statement from an international paediatric asthma consensus group. *Archives of Disease in Childhood*, 1992; **67:** 240–8.

Related topics of interest

Allergy (p. 17)
Cystic fibrosis (p. 104)
Lower respiratory tract infection (p. 211)
Upper respiratory tract infection (p. 319)

ATAXIA

Ataxia is the incoordination which results from sensory loss or cerebellar dysfunction. Presentation in childhood may be acute, intermittent or chronic. Clinical findings may include a wide-based gait, nystagmus, dysarthria and general hypotonia. There is difficulty in performing finely coordinated movements (handwriting, finger–nose test) and rapidly alternating movements (dysdiadochokinesia). A thorough history and neurodevelopmental assessment will indicate the likely aetiology in many cases. Specific investigations include CT and MRI to exclude space-occupying lesions of the head and spine, EEG, blood glucose and blood and urine analysis for toxicology and amino acids.

Acute ataxia

1. Drugs. Phenytoin, piperazine, thallium (in pesticides), alcohol and solvent abuse may all cause an acute ataxia.

2. Acute cerebellar ataxia of childhood. A dramatic onset of severe bilateral truncal ataxia occurs typically 1–3 weeks after a viral infection (echo, Coxsackie and influenza A and B have been implicated). It is usually seen in a young child of 1–2 years, with a notable absence of fever, altered consciousness or meningism. The cerebrospinal fluid (CSF) may show a mild pleocytosis and a later rise in protein. Investigation must always exclude a toxic or intracranial aetiology. Spontaneous recovery within 2 months is the rule.

3. Direct bacterial or viral invasion. Cerebellar abscess, meningitis or encephalitis (especially varicella) may cause ataxia. Fever, meningism, seizures and altered consciousness may also be prominent features.

4. Head injury. After direct trauma to the head, ataxia may continue for months.

5. Basilar artery migraine. Features of brain-stem dysfunction may dominate the clinical picture, and ataxia, dysarthria, tinnitus and vertigo may occur. Such attacks may be related to menstruation in adolescent girls, often with a strong family history of migraine.

6. Epilepsy. A post-ictal state or minor epileptic status, i.e. very frequent myoclonic seizures, may cause an acute ataxia.

7. Hysteria.

8. 'Dancing eye syndrome' (myoclonic encephalopathy of infancy). Chaotic, irregular and rapid jerking of the extraocular muscles and limbs is seen while the child is awake or asleep, simulating both nystagmus and ataxia. This syndrome often follows an upper respiratory tract infection, and there is also an association with neuroblastoma (often occult). The movements are present at rest and are not increased by movement. Eye movements have no constant relationship with the direction of gaze, unlike true cerebellar dysfunction. Initial improvement may occur with adrenocorticotrophic hormone (ACTH), but recurrent ataxia with neurodevelopmental problems is common.

Intermittent ataxia

Several of the inborn errors of metabolism may cause an intermittent ataxia, e.g maple syrup urine disease, arginosuccinic acidaemia. Drowsiness, vomiting and seizures may be associated features. Multiple sclerosis is uncommon in childhood but intermittent ataxia may be a presenting symptom.

Chronic ataxia

1. Ataxic cerebral palsy. Three main groups occur (distinction may be subtle): ataxic, ataxic diplegic and the dysequilibrium syndrome. These cases account for 10% of all cerebral palsy. Often these children have been term infants and the cerebral palsy is of diverse origin, e.g. genetic, brain maldevelopments (Arnold–Chiari) or post-natal insults such as encephalitis. The dysequilibrium syndrome is associated with an autosomal recessive inheritance, cerebellar hypoplasia, spasticity and mental retardation.

2. Hydrocephalus. Ataxia and abnormal pyramidal function may occur when hydrocephalus develops, with relatively few signs of raised intracranial pressure, e.g. aqueduct stenosis.

3. Cerebellar neoplasia. Medulloblastoma, astrocytoma, ependymoma and haemangioblastoma of the cerebellum often present with ataxia in addition to signs and symptoms of raised intracranial pressure.

4. Progressive degenerative neurometabolic disease. Ataxia is a common symptom in these diseases, especially in infantile and late infantile Batten's disease and metachromatic leucodystrophy. Seizures, visual failure and dementia contribute to a relentless progression of disease.

5. Friedrich's ataxia. A progressive ataxia of limbs and trunk occurs, with dysarthria, weakness and wasting and pyramidal tract dysfunction of the legs. There is loss of joint and position sense, absent or reduced tendon reflexes in the knee and ankle and extensor plantar responses. Scoliosis, pes cavus, optic atrophy, deafness and diabetes mellitus are associated findings. Cardiomyopathy may occur, sometimes leading to arrhythmias. Symptoms usually commence before the age of 15, and the gradual deterioration of neurological and cardiorespiratory function renders most patients chair-bound by 20–30 years of age. The disease is of autosomal recessive inheritance, and its gene has been localized to chromosome 9.

6. Ataxia telangiectasia. Progressive cerebellar ataxia appears in early childhood, and may be associated with oculomotor apraxia, mental retardation and, later, loss of joint and position sense and dementia. Telangiectasiae appear at the age of 3–5, notably on exposed areas of skin (ear pinnae, flexures of knees and elbows, face). One third of patients develop malignancy, e.g. lymphoma, sarcoma, and affected patients have poor humoral and cellular defences towards infection. Investigations show depressed levels of IgA and IgM with a raised level of α-fetoprotein. Death before adult life (recurrent infection and immobility) is common. Disease inheritance is autosomal recessive; the gene has been localized to 11q22–23.

Further reading

Brett EM. Ataxia. In: Brett EM, ed. *Paediatric Neurology,* 2nd edn. Edinburgh: Churchill Livingstone, 1991.

Related topics of interest

Cerebral palsy (p. 56)
Hydrocephalus (p. 174)
Malignancy in childhood (p. 217)

BEHAVIOUR

Many behavioural problems are variations of an expected pattern of development, especially during the toddler years, and parents may have unrealistic expectations of the child's performance. More serious behavioural problems arise if there is physical disability, developmental delay or if the child is deprived of love, affection and stability within the home environment.

History

It is helpful for parents to describe their perception of the problem, discussing its frequency and apparent precipitating factors. Reports from teachers or carers other than the parents may reveal that certain behaviour is displayed only for the parents. A full medical and developmental history is essential. Factors in the social history are often the most relevant, e.g. recent marital separation, family bereavement or birth of a sibling. An account of the problem should always be heard from the child if possible.

Assessment

It is important to know exactly what the parents expect from any interventional programme, and how consistently they are willing to contribute to this at home. Developmental delay should be excluded, and any obvious medical factors treated appropriately. The basis of most programmes is to avoid rewards or attention for the unwanted behaviour but provide reward for good behaviour, e.g. by praise, attention, stars on a chart.

Sleep problems

1. Settling and waking problems. The younger child may be difficult to settle off to sleep in the evening, or wake frequently during the night, demanding attention or drinks. A regular bedtime routine with a bath and winding down of activity is most likely to be successful. If the child then wakes regularly, rewarding the waking with drinks or play should be avoided. Sedatives should only be used for temporary intervention, in conjunction with a plan to modify behaviour.

2. Night-time attacks. Night terrors and sleep-walking occur during deep non-rapid eye movement sleep during the first third of the night. The child remains asleep during the episode, and has no recollection of events the next morning. Nightmares occur during rapid eye movement sleep, and the child wakes fully with vivid recollection of the dream. Night terrors, sleep-walking and nightmares usually remit spontaneously.

3. Insomnia. Depression, anxiety, or organic illness may cause insomnia.

Eating problems

Toddlers frequently refuse to eat, causing great anxiety for their parents. Most children will be thriving, and parents should be encouraged to offer meals regularly, ideally as part of a family meal-time, without coaxing, scolding or forcing. Snacks should be discouraged. General review of health and dietary advice is needed if there is failure to thrive.

Temper tantrums

These are common from 1 to 5 years of age during moments of intense frustration. Reinforcement of the behaviour occurs if the parents give in to the child's desires, and this should be avoided. Removing the child from the centre point of attention and into a corner or separate room may help this behaviour.

Hyperactivity

Many children are given the label hyperactive when their behaviour represents part of the normal spectrum of activity of early childhood. Hyperactivity implies that there is disorganized behaviour with little regard for other people or objects. Minimal brain dysfunction and developmental delay may be responsible in a small minority, and problems may persist into early adulthood. A few children may be truly sensitive to food dyes and additives, but they are not thought to be a major cause of hyperactivity in children.

Autism

Autism is a syndrome of abnormal behaviour which affects 2–4 children per 10 000. It is more common in boys than girls. Although many cases are idiopathic, a specific medical diagnosis can be made in a minority of children, e.g. tuberous sclerosis, fragile X. The essential features of autism are:

- Delayed and deviant language. Language delay may be severe, and 50% of affected children never develop useful language. Comprehensive and expressive language is abnormal, and there may be abnormal use of gestures.
- Impaired social interactions. Socially the child is characteristically aloof, with a poor attachment to parents, avoidance of gaze and a lack of display of affection. They may play alone with little regard for the activity of others.
- Repetitive and ritualistic behaviours. Purposeless actions may be repeated endlessly, with a lack of imaginative,

creative or symbolic play. Minor changes in the child's ritual or routine may produce great distress.

These features are usually prominent by 2–3 years of age. Abnormal behaviour is often present from early infancy, although in a few cases a period of normal development precedes the appearance of autistic features. Severe learning difficulties occur in 70% of children and are assessed by non-verbal IQ tests.

Management

Educational provision should be in a well-structured environment, coupled with a behavioural programme involving both school and home. Residential schooling may sometimes be more appropriate. Neuroleptic drugs may occasionally be of value.

Further reading

Stores G. Sleep problems. *Archives of Disease in Childhood,* 1992; **67:** 1420–1.

Wolff S. Childhood autism: its diagnosis, nature and treatment. *Archives of Disease in Childhood,* 1991; **66:** 737–41.

Related topics of interest

Developmental assessment (p. 111)
Developmental delay and regression (p. 115)
Hearing and speech (p. 165)

BIRTH INJURIES

The risk of trauma during labour and delivery has been reduced by improvements in antenatal care and obstetric practice. Predisposing factors include malpresentation (e.g. breech), precipitate delivery (e.g. grand multiparity, prematurity), cephalopelvic disproportion, instrumental delivery, twin delivery and macrosomia. Many injuries are unsightly but resolve quickly. A few have longer term implications.

Soft-tissue injuries

1. Bruising and oedema of the presenting part. This is very common and usually resolves rapidly. A caput succedaneum is a serosanguinous subcutaneous effusion over the presenting part of the head which subsides within 24 hours. Face presentation may cause particularly unsightly bruising and oedema. If presentation is breech, the buttocks, legs and scrotum may be very bruised, and these infants are particularly at risk of jaundice. The oedematous part of the scalp that is sucked into a Ventouse extractor is known as a chignon. Occasionally there is haemorrhage into the chignon or necrosis of the skin. A cephalhaematoma is a subperiosteal haemorrhage which most frequently occurs over the parietal bones (sometimes bilateral). The extent of the haemorrhage is limited by the suture lines and it usually presents as a fluctuant swelling on the second day of life. It is associated with a hairline fracture of the underlying bone and is more common following instrumental deliveries. No treatment is required and it resolves over 2–12 weeks, with a hard calcified rim sometimes giving the impression of a depressed skull fracture. Subaponeurotic haemorrhages are rare but are a cause of significant blood loss. Bleeding is not confined to a single cranial bone but spreads over the head and down towards the eyes.

2. Petechiae and subconjunctival haemorrhages. Facial petechiae and bruising are particularly common if the cord has been tight around the neck, and the face may appear cyanosed and bloated. Subconjunctival haemorrhages due to raised intrathoracic pressure are common following vaginal delivery.

3. Abrasions and skin punctures. Abrasions caused by forceps, scalp electrodes or fetal blood sampling occasionally become infected, resulting in abscess formation.

4. Incisions. Accidental scalpel incisions during Caesarean delivery may require suturing.

5. Subcutaneous fat necrosis. This may occur following pressure from forceps (e.g. over the zygoma) or as a result of prolonged pressure against the mother's pelvis *in utero*. It resolves spontaneously but occasionally there is residual scarring.

Fractures

Linear skull fractures are common but usually go undetected because complications such as subdural and extradural haemorrhage are rare. Vaginal delivery should be remembered as a possible cause of a skull fracture in the investigation of non-accidental injury during the neonatal period. Depressed skull fractures are uncommon and there is a risk of underlying brain compression. Traumatic intracranial haemorrhage and brain injury are rare, but severe compression of the skull can cause tearing of the tentorium cerebri and its accompanying vein, which results in severe haemorrhage and is frequently fatal.

Fractures of the clavicle are associated with shoulder dystocia or breech delivery. There may be an associated brachial plexus injury. The fracture heals with callus formation over 2–3 weeks. Fractures of other bones (e.g. ribs, long bones of the limbs) are less common.

Peripheral nerve injuries

Facial nerve palsy is usually unilateral and is associated with pressure from forceps or the maternal pelvis. The weakness generally resolves within days or weeks. Congenital defects of the seventh nerve are often bilateral.

Lateral flexion of the neck or traction on the arms during delivery may result in injury to the brachial plexus, phrenic nerve or cervical sympathetic nerves (Horner's syndrome). Injury to the upper brachial plexus roots (C5, C6) causes an Erb's palsy (waiter's tip position). The weakness usually resolves in a few weeks but may take several months. A small number of infants are left with residual weakness which may be improved with surgery. Damage to the lower brachial roots (C8, T1) occurs when the arms are extended up beside the head causing a Klumpke's palsy.

Spinal cord injuries

Fracture dislocation of the lower cervical spine is a very rare injury usually associated with breech delivery. There is flaccid quadriplegia and urinary retention, and the prognosis is very poor.

Intra-abdominal injuries Subcapsular haematomas of the liver and spleen are uncommon but may cause hypovolaemic shock and death. Predisposing factors include hepatosplenomegaly, coagulopathies, breech delivery and cardiac massage.

Further reading

Kliegman RM, Behrman RE. Birth injury. In: Kliegman RM, Behrman RE, eds. *Nelson Textbook of Pediatrics*, 14th edn. Philadelphia: WB Saunders, 1992; 453–8.

Related topic of interest

Neonatal jaundice (p. 231)

BLISTERING CONDITIONS

Blisters (bullae) are usually seen during childhood as a manifestation of acute trauma or infection, and complications are uncommon. However, bullae may also be primary lesions of certain dermatoses which are genetically inherited or immunologically mediated, and both morbidity and mortality may be significant in these groups.

Problems

- Mechanical.
- Secondary infection.
- Fluid and heat loss.
- Scarring.
- Associated medical problems.

Aetiology

1. Genetic

(a) Epidermolysis bullosa (EB)

- Junctional EB is a rare lethal autosomal recessive condition, presenting at or soon after birth and is characterized by a subepidermal split through the epidermal side of the basement membrane zone. There is severe skin and mucosal involvement.
- In EB simplex, there is a suprabasal split. Children tend to present in early childhood when they become mobile. Scarring is unusual, and mucous membranes are rarely affected. Inheritance is autosomal dominant.
- Dystrophic EB is characterized by a subepidermal blister below the junction of the dermis and epidermis. Disfiguring and disabling scarring may result. Autosomal dominant (mild) and recessive (severe) forms occur.

(b) Incontinentia pigmenti. In this X-linked dominant condition, linear vesicles are seen in the neonate, but these fade, and small papules later occur.

2. Physical. Excessive friction, heat or cold, ultraviolet light exposure or chemical irritation may cause blistering in previously normal skin.

3. Infection. Staphylococcal infection may produce blisters in the scalded skin syndrome (toxic epidermal necrolysis) which results from toxins of *Staphylococcus* phage type 71. Bullous impetigo may result from *Staphylococcus aureus* infection. Herpes zoster (chickenpox) is the commonest cause of a generalized vesicular eruption in childhood. Herpes simplex type 1 may cause clusters of small vesicles around the lips and face, and type 2 may be acquired by the neonate of an infected mother, causing a serious viraemia with scattered blisters. Vesicles on the tongue, hands and feet are seen with coxsackievirus A16 infection (hand, foot and mouth disease) as part of a mild, self-limiting illness.

4. *Inflammation.* Blisters occur in eczema, usually where the epidermis is thicker, such as palms and soles. Severe illness may arise if eczema is secondarily infected with herpes simplex – Kaposi's varicelliform eruption. Erythema multiforme is a condition which affects the skin and mucous membranes, usually after herpes simplex or mycoplasma infection, although drugs such as cotrimoxazole may be responsible. Target lesions are typical, a purple centre with red surround, but in severe cases the centre may be a vesicle or bulla. The severe form with mucosal involvement is Stevens–Johnson syndrome. Inflammation following insect bites or scabies may also produce bullae.

5. *Immunological.* Dermatitis herpetiformis occurs as grouped vesicles and papules over scapulae, knees, elbows and buttocks. Itching is often severe. Most patients have a gluten-sensitive enteropathy which may be asymptomatic.

Chronic bullous dermatosis of childhood presents with tense bullae on the skin of the lower half of the body, usually before 6 years of age. Itching may occur. Rosette arrangements of bullae are common, and the genitalia are often affected.

Bullous pemphigoid is a rare, non-itchy bullous condition occurring before the age of 6 years. Mucous membranes, face, genitals and limbs may be affected.

6. *Drugs.* Bullae occur rarely with drugs, e.g. fixed drug eruption with paracetamol, toxic epidermal necrolysis with non-steroidal anti-inflammatory drugs.

Clinical assessment

The age of onset of blistering is important, e.g. neonatal bullae formation with epidermolysis bullosa. There may be a clear history of eczema, infection, infestation, insect bites or exposure to physical agents or drugs. General examination may reveal a pyrexia, lymphadenopathy, poor growth or evidence of an enteropathy. The skin surrounding the bullae should be carefully assessed for evidence of excoriation, erythema or scar tissue. The distribution of the lesions should be noted, and in particular the mucous membranes and genital area must be examined.

Investigation

- Bacteriological and virological cultures if infection is suspected.
- Skin biopsy if genetic or immunological disease is suspected. One biopsy should be taken from the edge of a new blister (less than 24 hours old) for routine histology, and a second biopsy should be taken from uninvolved skin for immunofluorescence. Electron microscopy of the skin may also be of diagnostic help.
- Serum immunofluorescence. Circulating antibodies may be demonstrated by indirect immunofluorescence in the patient's serum in some diseases.

Management

Management depends on an accurate diagnosis. Infective causes should be treated with appropriate antibiotics or antiviral agents. Dermatitis herpetiformis responds to dapsone and a gluten-free diet. Chronic eczema may require topical steroid therapy, and oral corticosteroids may be needed for bullous pemphigoid. Unfortunately, no successful drug treatment is available for epidermolysis bullosa, and meticulous skin protection to lessen the scarring is the mainstay of care; genetic counselling should be offered.

Further reading

Verbov J, Morley N. *Colour Atlas of Pediatric Dermatology*. Lancaster: MTP Press, 1983.

Related topics of interest

Eczema (p. 127)
Rashes and naevi (p. 285)

BURNS

After road traffic accidents, burns are the commonest cause of accidental death in childhood in the UK. Many deaths occur in children under 5 years who are unable to make their own escape from house fires and die from smoke inhalation. Other children suffer burns from household heaters, open fires and cooking sources, or scalds from kettles, bath water or hot drinks, with considerable morbidity as a result of the injury and the cosmetic effects of scarring. Chemical and electrical burns are less common. Child abuse accounts for some burns.

Problems

- Pain.
- Respiratory distress.
- Scarring and contractures.
- Malnutrition.
- Acute renal failure.
- Sepsis.

The burn

At the site of an acute burn, there is loss of fluid from the plasma into the extracellular space, with loss of water, heat and protein. Multiple effects on immunity, metabolism and organ function follow the insult. A first-degree burn produces pain and erythema without blistering. A second-degree burn (partial thickness burn) destroys the epidermis but spares the dermis, and heals within 2 weeks. In a third-degree (full-thickness) burn, the dermis is also destroyed, the burn is white and painless, and healing occurs very slowly from the edges of the wound with contraction.

History

It is essential to establish the nature of the burn injury, and the time at which it occurred (or the duration of smoke inhalation). There may be a history suggestive of associated injuries, e.g. a fall or explosion. Brief enquiry should be made about tetanus immunization status and pre-existing illness.

Emergency management of major burns

Patients with over 10% of body surface area burned have major burns and the following aspects should be assessed:

- Resuscitation and cardiorespiratory support.
- Size and depth of the burnt area.
- Associated injuries.
- The need for transfer to a specialized burns unit.

Inhalational injury may be suggested by respiratory distress, facial burns, singed eyebrows and nasal hair or carbon deposits in the oropharynx. One hundred per cent oxygen should be given by facemask, and endotracheal intubation and intermittent positive-pressure ventilation should follow swiftly, before airway oedema occurs.If thoracic or limb burns are circumferential, escharotomies may be necessary.

An estimation of the burn surface area can be made. In patients over 12 years the rule of nines can be used. Below this age, the Lund and Browder chart, which corrects for the differential proportion of head and leg size related to age, may be used. A child's palm and closed fingers cover approximately 1% of the body surface area.

For major burns (above 10%), intravenous access should be established, through burnt skin if necessary, and preferably in an upper limb vein. Some children with burns of 5–10% may need intravenous fluid, and close observation is needed in this group of patients. Adequate analgesia is essential, and may be given as intravenous morphine (0.1 mg/kg) for major burns, or as oral paracetamol or intramuscular morphine for less severe burns. Antibiotics may be given, depending on local policy.

Investigations

- Weigh child if possible.
- FBC, haematocrit.
- Blood group and cross-match.
- U&E, urine specific gravity.
- Arterial blood gases, acid–base balance.
- Carboxyhaemoglobin.
- Chest radiograph; other radiographs as indicated.

Fluid replacement

This should be given as intravenous colloid, preferably human plasma protein fraction, and/or crystalloid. Replacement volumes are given in aliquots, depending on the size of the patient and the extent of the burn. A suitable formula for calculating the volume of fluid required in ml per time period = (total % of burn x weight in kg)/ 2 .

This volume of fluid is then given in the following six periods over 36 hours.

- 4 hours (from time of injury).
- 4 hours.
- 4 hours.
- 6 hours.
- 6 hours.
- 12 hours, i.e. now 36 hours from time of burn.

Although this formula of six periods of increasing length prescribes a defined amount of fluid until 36 hours post burn, the adequacy of resuscitation must be reassessed frequently, at least once during every period, and adjustments made as indicated by pulse, blood pressure, peripheral perfusion and urine output. In addition, maintenance fluid requirements may be given as oral fluids, or as intravenous 4% dextrose with 0.18% saline.

Further care

Following resuscitation, transfer to a specialist burns unit should be considered if over 10% of the body surface area is involved, if there are suspected full-thickness burns or if the site of the burn suggests a poor cosmetic or functional outcome.

Prevention

The production of flameproof furniture and nightclothes and a reduction in the use of open fires may help to prevent burns. The use of safety features such as fireguards, cooker guards, coiled flexes on kettles, and domestic smoke detectors should be actively encouraged. Smoking should be discouraged.

Further reading

Muir IFK, Barclay TL, Settle JAD. *Burns and their Treatment,* 3rd edn. London: Butterworths, 1987.

Parkhouse N. Burns and scalds. *Current Paediatrics,* 1993; **3:** 67-71.

Related topics of interest

Cardiopulmonary resuscitation (p. 52)

Shock (p. 291)

CALCIUM METABOLISM

Ninety-nine per cent of total body calcium is present in the skeleton. The other 1% in extracellular and intracellular fluids plays a vital role in enzyme reactions, coagulation and neuromuscular functioning. Vitamin D, parathyroid hormone (PTH) and calcitonin act on bone, kidneys and gastrointestinal tract to maintain serum calcium concentrations within narrow limits. Hypocalcaemia and hypercalcaemia may be associated with disorders of vitamin D metabolism, parathyroid disease or non-endocrine disorders such as chronic renal failure. Hypocalcaemia is a more common problem than hypercalcaemia in childhood and, apart from excessive ingestion of vitamin D, causes of hypercalcaemia are rare.

Calcium

Serum calcium is present in three fractions which are in dynamic equilibrium (50% ionized (active), 40% protein bound (inactive), 10% complexed to phosphate, citrate, etc.). Routinely, total serum calcium is measured, but ionized calcium can be measured directly. Acidosis increases ionized calcium (H^+ ions compete for albumin binding sites) and alkalosis decreases ionized calcium. Hypoproteinaemia may lead to a low total serum calcium while ionized calcium levels remain normal. Fasting blood taken with the least amount of venous stasis gives the most accurate level.

Vitamin D (calciferol)

This is the collective term for ergocalciferol (vitamin D_2) and cholecalciferol (vitamin D_3), which occur naturally in foods, e.g. fish, eggs, butter. Absorption via the upper small intestine is impaired by steatorrhoea. Vitamin D_3 is also produced in skin by UV light acting on a precursor. Vitamin D is activated by hydroxylation in the liver to 25-hydroxy D and then in the proximal renal tubules to 1,25-dihydroxy D.

Actions
Increases serum calcium by:

- Increasing intestinal absorption.
- Increasing mobilization from bone (PTH dependent).
- Decreasing renal excretion.

Parathyroid hormone (PTH)

PTH is an 84 amino acid peptide secreted by the chief cells of the parathyroid glands. It is coded for by a gene close to that for insulin on chromosome 11. PTH is cleaved from larger, biologically inactive precursors and released into the circulation with carboxy-terminal fragments. It is metabolized predominantly by the liver. There is diurnal variation in PTH secretion with higher levels in the early morning. Secretion is increased by hypocalcaemia, catecholamines, vitamin D metabolites and cortisol, and is suppressed by hypercalcaemia.

Actions

(a) Increases serum calcium by:

- Increasing resorption from bone.
- Decreasing renal excretion.
- Stimulating renal 1,25-dihydroxy D synthesis.

(b) Increases renal excretion of phosphate and bicarbonate.

Calcitonin

Calcitonin is a 32 amino acid peptide synthesized and secreted by the parafollicular (C) cells of the thyroid gland, encoded by a gene on chromosome 11. Secretion is suppressed by hypocalcaemia, and increased by hypercalcaemia, gastrointestinal hormones, oestrogens and β-adrenergic agonists.

Actions

- Inhibits calcium resorption from bone.
- Increases renal excretion of calcium, phosphate, magnesium, sodium, potassium.
- Promotes effect of vitamin D on intestinal absorption of calcium.

The importance of calcitonin in homeostasis remains unclear as neither thyroidectomy nor calcitonin hypersecretion affect serum calcium levels. It may protect against unwanted bone resorption when demands for calcium are high, e.g. pregnancy, lactation.

Hypocalcaemia

Aetiology

- Hypoparathyroidism (decreased secretion = true hypoparathyroidism; decreased peripheral action of PTH = pseudohypoparathyroidism).
- Vitamin D deficiency – lack of sunshine, dietary deficiency, malabsorption.
- Acute or chronic renal failure.
- Renal tubular disorders – Fanconi syndromes.
- Hypomagnesaemia – i.v. feeding, drugs, e.g. cisplatin.
- Long-term anticonvulsant treatment (phenytoin).
- Citrated blood.

Clinical features

If the onset is rapid there will be neuromuscular manifestations such as paraesthesia around the mouth, hands and feet, cramps, carpopedal spasms and tetany (Trousseau's and Chvostek's signs). Convulsions may occur, particularly in infancy. Neonatal hypocalcaemia is a relatively common transient problem. Chronic hypocalcaemia results from vitamin D deficiency or hypoparathyroidism.

Rickets is due to defective growth plate mineralization secondary to decreased calcium or phosphate availability:

(a) Calciopenic rickets:
- Vitamin D deficiency.
- Defective conversion of vitamin D to 1,25-dihydroxy D (e.g. renal failure).

(b) Phosphopenic rickets:
- Renal phosphate loss (Fanconi syndrome, vitamin D-resistant rickets).
- Inadequate phosphate intake (premature infants).

Clinical features of rickets include craniotabes, delayed closure of the anterior fontanelle and swelling of the metaphyses (particularly the wrists). Later features include softening and deformity of long bones, chest deformity and enlargement of the rib ends (rickety rosary).

Hypoparathyroidism may be sporadic or familial, transient or permanent. It may be idiopathic or secondary (e.g. neck surgery or irradiation). It may be associated with mucocutaneous candidiasis, Addison's disease and other autoimmune diseases. Chronic hypocalcaemia leads to ectodermal changes (dry skin, coarse hair, brittle nails, tooth enamel hypoplasia), cataracts, basal ganglia calcification and mental retardation.

Investigation

- Ionized calcium.
- Phosphate, alkaline phosphatase. Low phosphate with raised alkaline phosphatase suggests rickets. Phosphate will be raised in both hypoparathyroidism and pseudo-hypoparathyroidism.
- Total protein/albumin.
- Serum U&E, creatinine and urinalysis to exclude a primary renal cause.
- Serum magnesium. Correction of low magnesium may correct hypocalcaemia.
- Serum PTH. Low or undetectable in hypoparathyroidism, high in pseudohypoparathyroidism and rickets.
- Urinary cAMP and phosphate increase in response to PTH in hypoparathyroidism but not in pseudo-hypoparathyroidism.

- Plasma vitamin D metabolites. Low levels found in dietary rickets, liver disease, malabsorption and anticonvulsant treatment.

Management

1. Acute. Intravenous calcium gluconate infusion.

2. Long-term. Depends on cause, e.g. vitamin D or phosphate supplements.

Hypercalcaemia

Aetiology

- Primary hyperparathyroidism (parathyroid hyperplasia, adenoma).
- Hypervitaminosis D.

(a)Increased intake (e.g. overtreatment of hypoparathyroidism, rickets, etc.)
(b) Increased synthesis (increased conversion of 25-hydroxy D to 1,25-dihydroxy D seen in some patients with sarcoidosis, tuberculosis or malignancy).

- Familial hypocalciuric hypercalcaemia (due to defective renal excretion of calcium).

Clinical features

Features include nausea and vomiting, constipation and polyuria. Irritability and failure to thrive are common in infants. Renal stones, bone pain and pathological fractures may occur. Nephrocalcinosis and ectopic calcification will occur if hypercalcaemia is long-standing, with subsequent renal failure. Idiopathic hypercalcaemia of infancy associated with mental retardation, cardiovascular abnormalities and dysmorphic features is known as William's syndrome. Primary hyperparathyroidism may be familial and is associated with the multiple endocrine neoplasia syndromes (MEN 1 and 2).

Investigation

- Phosphate, alkaline phosphatase.
- Total protein/albumin.
- Serum PTH. Inappropriately normal or raised in presence of hypercalcaemia indicates hyperparathyroidism.

Management

1. Acute. Steroids and i.v. hydration ± calcitonin.

2. Long-term. Remove underlying cause.

Further reading

Bishop N. Bone disease in preterm infants. *Archives of Disease in Childhood,* 1989; **64**: 1403–9.

Kruse K. Disorders of calcium and bone metabolism. In: Brook CGD, ed. *Clinical Paediatric Endocrinology.* 2nd edn. Oxford: Blackwell Scientific Publications, 1989: 508–49.

Related topics of interest

Chronic renal failure (p. 75)
Polyuria and renal tubular disorders (p. 267)

CARDIOPULMONARY RESUSCITATION (CPR)

Cardiac arrest in childhood is usually secondary to hypoxia and is rarely due to primary cardiac disease. The outcome for respiratory arrest alone is often good, but the outcome for cardiac arrest is generally poor, even in children who are successfully resuscitated, as tissue hypoxia and acidosis prior to arrest frequently lead to subsequent multisystem failure. Following successful CPR, management aims to achieve and maintain homeostasis to optimize chances of recovery. Advanced paediatric life support (APLS) courses are run regularly in the UK to help doctors, nurses and paramedics deal with seriously ill and injured children more effectively. Prevention of cardiac arrest by earlier recognition of a seriously ill child will help reduce morbidity and mortality. The current CPR recommendations of the Advanced Life Support Group are given here.

Aetiology

1. Respiratory failure

- Respiratory distress, e.g. asphyxia, epiglottitis, asthma, bronchiolitis.
- Respiratory depression, e.g. raised intracranial pressure, convulsions, poisons.

2. Circulatory failure – shock

- Hypovolaemic.
- Cardiogenic.
- Septicaemic.
- Anaphylactic.

Basic life support

1. Call for assistance, approach the child with care and remove from continuing danger.

2. Airway. If the child is unresponsive to the question "Are you alright?" or gentle shaking, open and maintain the airway using the head tilt/chin lift manoeuvre. The jaw thrust manoeuvre is safer if there may be a cervical spine injury.

3. Breathing. Look, listen and feel for signs of breathing. If opening the airway has not led to resumption of breathing give five long mouth to mouth/nose breaths.

4. Circulation. Feel for a central pulse for 5 seconds (brachial or femoral in infants, carotid in older children). If absent commence cardiac compression at a rate of 100 per minute. A precordial thump is not recommended in children. Cardiac compression in infants is achieved by using two fingers to depress the sternum by 1.5–2.5 cm at a point one finger-breadth below the nipple line. Alternatively, the infant is held with both the rescuer's hands encircling the chest so the thumbs can compress the sternum in the correct position. In small children the heel of one hand is used to compress the sternum one finger-breadth above the xiphisternum to a depth of 2.5–3.5 cm. In larger children the heels of both hands are used to compress the sternum two finger-breadths above the xiphisternum to a depth of 3–4.5 cm. A ratio of five compressions to one breath should be maintained (20 CPR cycles per minute).

Advanced life support

1. Airway and breathing. Look, listen, and feel for airway obstruction, respiratory arrest, depression or distress. If there is respiratory arrest or depression despite airway opening manoeuvres, commence ventilation with high-concentration oxygen via bag and mask or orotracheal intubation. Uncuffed tubes are preferred up to puberty as they cause less oedema at the cricoid ring. If intubation is impossible or unsuccessful consider needle cricothyroidectomy. Look for evidence of chest or neck trauma, and assess symmetry of chest movement and breath sounds.

2. Circulation. If there are no palpable central pulses, cardiac compression should be continued at a rate of 100 per minute, not stopping while a breath is given. Establish venous access. Commence pulse oximetry and ECG monitoring.

3. Arrhythmias. The child should have a secure airway and be adequately ventilated with high-concentration oxygen before any drugs are given. For children aged 1–10 years the weight (kg) can be estimated as 2 x (age + 4).

Arrhythmias

Asystole

Asystole is the most common arrhythmia as prolonged severe hypoxia and acidosis cause progressive bradycardia leading to asystole.

- Adrenaline 10 μg/kg i.v. or i.o. (intraosseous) or 100 μg/kg via an endotracheal tube (ETT) (10 μg = 0.1 ml of 1:10 000).
- Atropine 20 μg/kg i.v. or i.o.
- Sodium bicarbonate 1 mmol/kg i.v. or i.o. It may not be given via an ETT. Further doses should be given as indicated by the venous pH.

- Adrenaline 100 µg/kg i.v. or i.o. Further doses given every 3 minutes (i.e. 60 cycles x 5:1 CPR cycles).
- Consider volume expansion.

Ventricular fibrillation (VF)

Ventricular fibrillation (VF) is uncommon but may occur in those with cardiac disease or recovering from hypothermia, or poisoned with tricyclics.

- Defibrillate 2 J/kg.
- Defibrillate 2 J/kg.
- Defibrillate 4 J/kg.
- Adrenaline 10 µg/kg i.v. or i.o.
- Three further shocks (4 J/kg).
- Adrenaline 100 µg/kg i.v. or i.o.
- Three further shocks (4 J/kg) 1 minute later.

VF due to hypothermia may be resistant until the core temperature is increased. Antiarrhythmic agents (e.g. lignocaine) and sodium bicarbonate should be considered in drug overdose.

Electromechanical dissociation

No output in the presence of QRS complexes on the ECG most commonly occurs in profound shock. It may also occur with tension pneumothorax, cardiac tamponade, drug overdose, hypothermia and electrolyte imbalance.

- Adrenaline 10 µg/kg i.v. or i.o., or 100 µg/kg via an ETT.
- Volume expansion with 20 ml/kg crystalloid i.v. or i.o.
- Look for the underlying cause and treat appropriately.
- Adrenaline 100 µg/kg i.v. or i.o. every 3 minutes (i.e. 60 x 5:1 CPR cycles).

Post-resuscitation management

1. Monitor blood pressure, heart rate and rhythm, oxygen saturation, arterial pH and gases, toe–core temperature and urine output. CVP monitoring is often useful. The role of intracranial pressure (ICP) monitoring is controversial.

2. Investigations. Chest radiograph, FBC (haematocrit and platelets), group and save serum for cross-match, U&Es, clotting screen (DIC, hepatocellular damage), blood glucose, liver function tests, ECG.

3. Maintain oxygenation.

4. Maintain circulation. Normalize acid–base and electrolyte balance and correct hypovolaemia. Inotropic support may be needed.

5. Normalize blood glucose and body temperature.

6. Maintain adequate analgesia and sedation. Detect and treat raised ICP.

Stopping resuscitation

CPR should be discontinued if there is no detectable cardiac output or cerebral activity after 30 minutes. Prolonged resuscitation attempts are indicated in the hypothermic child (continue until core temperature is at least 32°C) or those who have been poisoned with central depressant drugs.

Further reading

Life support. In: Advanced Life Support Group. *Advanced Paediatric Life Support – the Practical Approach.* London: BMJ Publications, 1993: 21–58.

Related topics of interest

Coma (p. 85)
Shock (p. 291)

CEREBRAL PALSY

Cerebral palsy has been defined as a persistent but not unchanging disorder of movement and posture due to a non-progressive disorder of the immature brain. The estimated prevalence is 2 per 1000 live births.

Problems

- Mobility.
- Contractures and deformities.
- Vision, hearing and speech.
- Epilepsy.
- Nutrition.
- Education and employment.
- Social.

Aetiology and classification

Aetiological factors may be prenatal, e.g. congenital infection; perinatal, e.g. placental abruption; or post-natal, e.g. head injury. However, no clear aetiological factor can be identified in two-thirds of cases. Classification is based on the predominant disorders of movement and posture, e.g. spastic diplegia, athetoid (dyskinetic), etc., but the changing expression of the brain insult over time means that such classification must remain flexible.

Congenital hemiplegia

Infants may present between 3 and 9 months with paucity of movement on the affected side and asymmetry of tone. The arm is more severely affected than the leg, with distal weakness, and vasomotor changes may be seen. Motor milestones are delayed. There may be a homonymous hemianopia, fine sensory defects and cranial nerve involvement.

Spastic diplegia

Limbs on both sides of the body are affected, but lower limb involvement dominates. There is a high proportion of preterm births in this group (periventricular leucomalacia or haemorrhage is often found in association). The baby may initially be hypotonic, then there is a generalized increase in tone, with 'scissoring' of the legs, persistence of primitive reflexes and the finding of brisk tendon reflexes in the legs.

Spastic quadriplegia

This group includes the most severely affected children, since both upper and lower limbs are spastic, and mental retardation, pseudobulbar palsy, visual problems and epilepsy are common associated features. Primitive reflexes often persist, e.g. tonic neck reflexes.

Dyskinetic (athetoid) cerebral palsy

The affected neonate is often hypotonic with poor feeding and seizures, but later choreoathetoid movements with spasticity develop. Hypoxic ischaemic damage and neonatal hyperbilirubinaemia are major aetiological factors. High-frequency deafness is a common finding in dyskinetic cerebral palsy due to neonatal jaundice.

Ataxic cerebral palsy

Athough perinatal risk factors may be present in this group, there is also a high incidence of genetic factors, hydrocephalus and structural abnormalities of the cerebellum. Ataxia may be pure, or occur as a mixed form – ataxic–diplegia. Early hypotonia may be noted, followed by increasing spasticity, an intention tremor and truncal ataxia.

Diagnosis and assessment

In many cases, cerebral palsy will be diagnosed during the developmental surveillance of a child with known risk factors, e.g. extreme prematurity, neonatal meningitis. Early diagnostic confusion may arise with ataxic or dyskinetic forms, because progressive neurological or neuromuscular diseases may have similar clinical features. Observation of the child's movements and posture while at rest and during the stress of motor tasks may be particularly informative. Primitive reflexes are frequently retained. A full neurological examination including cranial nerves, optic fundi, tone and reflexes should be performed. Vision and hearing deficits should be formally excluded. Repeated assessments after diagnosis ensure appropriate planning of medical, social and educational services.

Investigation

Specific investigations are not routinely indicated for diagnosis, although cerebral imaging may show an anatomical lesion in some cases.

Management

The child with cerebral palsy requires the skills of many professionals – a multidisciplinary approach. Such a team may be led by a developmental paediatrician, and usually includes the skills of a physiotherapist, ophthalmologist, speech therapist, audiologist, teacher, ear, nose and throat (ENT) and orthopaedic surgeon, occupational therapist, etc. Specific management issues include:

1. Motor problems. Physiotherapy has a central role in the prevention of deformity and the promotion of more normal movement and posture, aiming for independent mobility whenever possible. Adaptive seating, chairs, standing frames and other aids may be employed. Various methods of physiotherapy exist, although not all methods will be appropriate for every child or therapist. The Bobath method guides the child through the normal motor milestones, positioning the child so that deformities are minimized. The controversial Peto technique involves one person acting as teacher and physiotherapist, usually in a residential setting.

2. Orthopaedic problems. Functional improvements may occur after elongation of the Achilles tendons or a wrist arthrodesis. Deformities such as hip dislocation and scoliosis may require surgical correction. Laboratory studies of gait analysis may now be used to select appropriate surgical candidates.

3. Feeding and speech. Oromotor dysfunction is common, with problems of drooling, poor feeding and dysphagia. Early speech therapy is important. Nutritional support may be needed by nasogastric tube or gastrostomy feeding. Gastro-oesophageal reflux is common, and difficult to treat.

4. Education. A pre-school centre which has facilities for the child's special needs provides a base from which a decision about the school placement can be made. This will follow a statement of educational needs, based on reports by all the involved professionals, and with the agreement of the parents on its conclusions.

Further reading

Hagberg B, Hagberg G. The origins of cerebral palsy. In: David TJ, ed. *Recent Advances in Paediatrics,* Vol 11. Edinburgh: Churchill Livingstone, 1993: 67–83.

Related topics of interest

Ataxia (p. 32)
Developmental delay and regression (p. 115)
Seizures (p. 288)

CHILD ABUSE

The true prevalence of child abuse is difficult to establish, as much abuse is minor and goes undetected. Young children under 2 years are most at risk from severe physical abuse, and first-born children are more commonly affected. Marital instability, young mothers, alcohol, poor housing and unemployment are all risk factors for child abuse. Frequently the abuser is a cohabitant but not the child's parent. Adults are much more likely to abuse if they were abused during their own childhood.

Presentation of abuse

1. Bruising. Accidental bruises are common on shins and foreheads, but bruising on the face, in the ears, on the shoulders, buttocks or back is suggestive of abuse. Patterns of fingertips or objects may be seen. Bruising of the thighs or genitalia should raise the suspicion of child sexual abuse. Superficial bruising is red/purple within 24 hours of injury, purplish blue at 24–48 hours, brown at 48–72 hours and yellow to fading thereafter.

2. Fractures. Fractures are difficult to detect clinically in the under-3-year age group, but may present with excessive crying, swelling or reluctance to use the affected limb. A skeletal survey is therefore indicated in younger children and infants. Multiple fractures, fractures of different ages, rib fractures (caused by squeezing), spiral, metaphyseal or epiphyseal fractures of long bones (pulling and twisting) and wide skull fractures are more likely to be caused non-accidentally than others. Differential diagnoses include birth trauma, osteogenesis imperfecta, copper deficiency and pathological fractures.

3. Burns and scalds. Non-accidental burns and scalds may occur by various methods, e.g. cigarette-tip burns, scalding bathwater or holding part of the child directly on a heat source. Blistering of the lesion may cause diagnostic confusion with staphylococcal bullous impetigo.

4. Child sexual abuse. Behavioural changes, a disclosure by the child or local symptoms of vulval or anal soreness may lead to suspicion of the diagnosis. Frequently there are no specific physical signs. Perianal bruising, fissures, hymenal tears and abrasions may be seen. Reflex anal dilation is a pointer to sexual abuse but is also seen in severe constipation, and is unreliable as a single physical sign.

5. Emotional abuse and neglect. This may present as failure to thrive, developmental delay, withdrawn or aggressive behaviour, poor school work, frequent accidents and a poor standard of cleanliness and care.

6. Munchausen syndrome by proxy. Various symptoms and signs are fabricated to bring the child to the attention of medical staff, e.g. seizures, blood in urine or vomit. The mother is usually psychiatrically disturbed, and may have previous close experience of medical or nursing matters.

7. Smothering. Episodes of apnoea or post-anoxic seizure may be the result of smothering with a pillow against the infant's face. There may be no clinical signs of injury. Cardiorespiratory disease, gastro-oesophageal reflux, seizures and biochemical disturbances must be excluded.

8. Poisoning. Children may present with a story that they have ingested the drug accidentally, or may present with episodes of unexplained symptoms, e.g. thirst after poisoning with table salt.

Clinical assessment

The history and examination should be conducted in the normal way, noting precisely the accounts given for the injuries, and by whom. Delayed presentation of injuries and a changing story of the mechanism and timing of injury should arouse the suspicion of non-accidental injury (NAI). A full social history should be taken, including details of all carers involved with the child. The child's height and weight should be documented, and behaviour, affect and developmental stage noted. Any injuries should be carefully recorded on a topographical chart, measuring bruises and indicating their colour. Retinal haemorrhages are easier to view after mydriatic drops in infants. In suspected sexual abuse, repeated examinations should be avoided, and examination of the external genitalia should be performed at the end of the examination, taking suitable swabs at this time.

Investigation

- FBC, coagulation screen. Exclude underlying blood dyscrasia in children with bruising.
- Skeletal survey. Periosteal new bone formation is the earliest sign, at 10–14 days after fracture, then callus from 14 days. Radioisotope bone scan may detect recent fractures in the presence of a normal skeletal survey.
- Serum copper and bone biochemistry if there are fractures.

- Medical photographs.
- Microbiological and forensic swabs in suspected sexual abuse.
- CT or MRI scan of head if intracranial haemorrhage is suspected.

Management

Immediate medical treatment of injuries may be needed, e.g. head injuries, fractures. The emergency protection of the child is the next priority, and should be arranged in consultation with social workers and police. Admission to hospital is often appropriate, allowing time for medical investigations, and may be on a voluntary basis, although legal action is sometimes needed. A case conference is arranged which allows sharing of information about the incident and the family from all professional agencies, and formulates a child protection plan. This may include placing the child's name on the child protection register.

Further reading

Chapman S. The radiological dating of injuries. *Archives of Disease in Childhood,* 1992; **67**: 1063–5.

Meadow R. *ABC of Child Abuse,* 2nd edn. London: BMJ Books, 1993.

Royal College of Physicians. *Physical Signs of Sexual Abuse in Children 1991.* London: Royal College of Physicians, 1991.

Related topic of interest

Children and the law (p. 65)

CHILDHOOD EXANTHEMATA

Problems

- Acute febrile illness.
- Congenital infection.
- Arthritis.
- Neurological sequelae.

Measles

Measles is caused by an RNA virus, spread by respiratory droplet. Its frequency has declined in recent years, since the introduction of measles vaccine in 1968 and the measles, mumps, rubella (MMR) vaccine in 1988, but it remains a significant cause of mortality and morbidity in developing countries, where malnutrition predisposes children to a more severe illness. The incubation period is 10 days. There is a prodromal illness of fever, conjunctivitis, coryza, with a dry cough. Lymphadenopathy and Koplik's spots (tiny white spots on a bright-red buccal mucosa) may be found. After 3–4 days, a florid maculopapular rash starts to appear on the head and neck, spreading down to cover the whole body, and this fades over 3–4 days. The child is infectious from the start of the prodrome to 4 days after the appearance of the rash. Complications are otitis media, pneumonia encephalitis and, rarely, the late complication of subacute sclerosing panencephalitis. Human normal immunoglobulin should be given to immunosuppressed children who have been in contact with measles.

Rubella

Rubella (German measles) is a mild infectious disease which usually affects children of 4–9 years. The incubation period is 14–21 days. There may be fever, malaise, a transient pink macular rash and cervical, subauricular and occipital lymphadenopathy. Such features may be diagnosed as a range of 'viral illnesses', and therefore a clinical diagnosis of rubella is unreliable, and is not a contraindication to vaccination. The child is infectious from 1 week before the rash appears to 4 days afterwards. The congenital rubella syndrome (CRS) may result from maternal rubella infection during the first 8–10 weeks of pregnancy, causing intrauterine growth retardation, deafness and cardiac defects. MMR vaccine has been given routinely to 12- to 18-month old children since October 1988. Unimmunized girls between 10 and 14 years and non-pregnant seronegative women of child-bearing age should be immunized.

Varicella

Varicella (chickenpox) is a highly infectious disease, most common in younger children in the spring. It is highly infectious, spreading readily by droplet or personal contact, especially within families. The incubation period is 14 days, and the child is infectious from 2 days before the rash appears until the vesicles are dry. There is no prodromal illness, and vesicles appear on the face and scalp in crops, spreading to the trunk, abdomen and limbs, often with intense pruritus. There may be vesicles in the mouth. Complications are secondary infection of the lesions and encephalitis (usually mild, with ataxia) at 7–10 days after the rash. In ill or immunosuppressed adults, pregnant women and neonates, morbidity and mortality are significant, featuring disseminated or haemorrhagic disease and severe varicella pneumonia.

Erythema infectiosum

Human parvovirus B19 is now known to be responsible for this acute illness in children, also known as fifth disease or slapped cheek disease. A mild febrile illness is accompanied by a reticular rash and bright-red cheeks bilaterally. Arthralgia and arthritis, usually in the small joints of the hands, occur in 10% of children. Intrauterine infection may result in spontaneous abortion and fetal hydrops, and aplastic crises may occur in those with an underlying haemolytic disorder.

Roseola infantum

Human herpes virus type 6 is responsible for this short febrile illness. As the fever settles, an erythematous macular rash appears. This virus has been implicated in various other clinical syndromes.

Staphylococcal toxic shock syndrome

The staphylococcal toxic shock syndrome is caused by an enterotoxin produced by the staphylococcus under certain conditions. It usually presents with a rapid onset of fever with delirium, profuse diarrhoea, erythematous rash, strawberry tongue, mucous membrane ulceration, shock and multiorgan system failure. There may be no focus of staphylococcal infection, carriage in the nasopharynx or an obvious source such as an abrasion or surgical wound. Rapid and aggressive resuscitation with intensive care is needed, combined with effective treatment of the staphylococcus and its source. A similar profound illness may occur with toxin-producing streptococci.

Lyme disease

A tick-borne spirochaete, *Borrelia burgdorferi,* is responsible for this illness. After the bite of an infected tick, a skin lesion called erythema chronicum migrans develops, initially as a red maculopapule, which then spreads circumferentially, although secondary lesions distant from the bite may occur. There is often central fading of erythema. Initially, other symptoms are fever, malaise and lymphadenopathy, but later arthritis, neurological and cardiac sequelae may occur. Treatment is with penicillin, erythromycin or cephalosporins.

Further reading

Cryan B, Wright DJM. Lyme disease in paediatrics. *Archives of Disease in Childhood,* 1991; **66**: 1359–63.

Related topics of interest

Congenital infection (p. 98)
Immunization (p. 184)
Pyrexia of unknown origin (p. 282)

CHILDREN AND THE LAW

Laws regarding children are of particular importance in issues of child protection, but they may also concern aspects of education, adoption and child safety. The most significant piece of legislation in recent years has been the Children Act 1989, a wide-reaching reform of many previous acts of parliament regarding children, and of related court procedures.

The Children Act 1989

The Children Act 1989 was implemented in October 1991. Its aims are to restructure the court proceedings regarding child protection, providing a unified and practical approach for all agencies involved.

The General principles of the Children Act 1989 are:

- The children's welfare is of paramount importance.
- The children should be brought up with their families whenever possible. Local authorities should promote this, provided it is consistent with the children's welfare.
- Delay in court proceedings for child protection issues is harmful and should be avoided whenever possible.
- Courts should refer to a checklist of matters regarding children's welfare.
- Court orders should not be made unless they positively contribute to children's welfare.
- The concept of parental responsibility (replacing parental rights) is retained even if children become the subject of a care order.
- Children should give their own consent to medical investigation and treatment, if they have sufficient understanding.

Emergency protection procedures

Voluntary agreements with parents may allow a period of assessment of risk to the child, without the need for a court order, but in some cases access may be denied, or the child may be removed from the place of safety, and specific court actions may be required.

- The Emergency Protection Order (EPO) may be made in cases of immediate physical risk. The order lasts for up to 8 days, but may be challenged after 72 hours. The courts may extend it by a further 7 days, but only once. Applicants may be Social Services officers or the NSPCC. Police also have powers to remove the child into police protection for 72 hours if it is believed that the child may otherwise suffer significant harm.

- A Child Assessment Order (CAO) lasts for 7 days and enables an assessment of the child's health and welfare to take place in cases of suspicion of significant harm, often overriding the objections of the parents. However the CAO may not override the wishes of the child, if he or she is of sufficient understanding to make an informed decision. The local authority or the NSPCC may apply.

Care proceedings (either a Care Order or a Supervision Order) may be required if the child is suffering, or likely to suffer, significant harm.

- A Care Order places the parental responsibility with Social Services (in addition to the parents), and usually removes the child from home to foster homes, or a children's home, at least temporarily. There is usually a contract between Social Services and the parents, following full multidisciplinary review at a case conference.
- A Supervision Order gives Social Services the right to visit a child and impose conditions relevant to the child's management, e.g. nursery or clinic attendance.
- Interim Care Orders may be made, placing the child in local authority care while a full investigation of the child's risk and needs is undertaken.

Private law orders

The court can decide also to make private law orders concerning the child's welfare (the four Section 8 orders).

- A Residence Order determines with whom the child is to live.
- A Contact Order decides what contact the child may have with other people.
- A Specific Issues Order is required for any specific care issue for an individual child, e.g medical care or consent issues.
- A Prohibited Steps Order may forbid any specific step which may be taken by the parents which is not in the interests of the child, e.g. unsupervised visiting.

The law and education

The Education Act 1944

This Act stated that every child over 5 years old must receive full-time education until the end of the spring or summer term following the child's 16th birthday.

The Education Act 1981

This followed the Warnock report of 1978 recommending that "all children should have education appropriate to their needs", aiming to educate children in normal schools whenever possible. If it is known by health authority staff that a child above 2 years old is likely to have special educational needs, staff should inform the Local Education Authority (LEA) after obtaining permission from the parents. The LEA should then obtain written information from all professionals involved with the child before preparing a Statement of Educational Need, which is sent to the parents, and a school most appropriate to those needs is then nominated by the LEA. Parents have 30 days in which to appeal against the statement. Statements should be reviewed annually.

Further reading

Leslie SC. Child protection – The Children Act 1989. *Current Paediatrics,* 1993; **3(3):** 168–70.

Working Together Under the Children Act 1989. London: HMSO, 1991.

Related topic of interest

Child abuse (p. 59)

CHROMOSOMAL ABNORMALITIES

Abnormalities of chromosomal number or structure result in characteristic syndromes of multiple congenital malformations. These are usually associated with learning difficulties. A chromosomal abnormality is found in 25–30% of early abortions and in 5–10% of stillbirths and infants dying in the first year. Ultrasound scanning may detect abnormalities antenatally and the diagnosis is confirmed by amniocentesis or chorionic villus biopsy.

Down's syndrome (trisomy 21)

Down's syndrome is the commonest autosomal trisomy, affecting 1 in 650 live births, with the same number of fetuses failing to reach term. Most cases are due to non-dysjunction of chromosome 21 during meiosis. Only 5% of cases are due to translocation, and mosaics (some cells trisomy 21, some normal 46 chromosomes) account for only 2% of the total. The risk of having an affected child increases with maternal age, but 50% of affected infants are born to mothers under 35 years because of the higher birth rate in this age group. The overall risk is 1 in 650, but at 20 years it is only 1 in 2000. This rises steeply with maternal age to 1 in 450 risk at 35 years, 1 in 100 at 40 years and 1 in 30 at 45 years. The recurrence risk following one child with Down's syndrome is 1% if due to non-dysjunction but 10% if the mother carries a translocation (usually on to chromosome 14). The recurrence risk if the father is the translocation carrier is 2–3%. Amniocentesis should be offered to women with increased risk, i.e. age over 35 years, previous affected child, known balanced translocation carrier.

Clinical features

- Characteristic facial features. Epicanthic folds, flat occiput, protruding tongue.
- Eyes. Brushfield spots, nystagmus, myopia, early cataracts.
- Hands and feet. Brachydactyly, clinodactyly (incurved little fingers), abnormal dermatoglyphics, sandal toes (wide gap between first and second toes).
- Hypotonia. Motor delay, feeding difficulties, recurrent chest infections, increased risk of congenital dislocation of the hip and hernias.
- Gastrointestinal abnormalities. Increased risk of atresias (especially duodenal) and Hirschsprung's disease.
- Congenital heart defects (50%). Particularly septal defects, Fallot's tetralogy.
- Learning difficulties. Usually mild or moderate but rarely severe. High incidence of Alzheimer's disease in third and fourth decades.
- Glue ear. High incidence of conductive deafness.
- Leukaemia. Increased risk.

Edward's syndrome (trisomy 18)

This is the second commonest autosomal trisomy but the prevalence is only 1 in 4000 births. It is more common in girls than boys. The majority die in the neonatal period but a few have survived several years. It is associated with increased maternal age but the recurrence risk is low.

Clinical features

- Polyhydramnios and intrauterine growth retardation (IUGR).
- Craniofacial abnormalities. Low-set ears, prominent occiput, narrow head, micrognathia, narrow slanting palpebral fissures.
- Limb abnormalities. Clenched hands with overlapping fingers, rocker bottom feet, hypoplastic nails, flexion deformities.
- Congenital heart defects.
- Gastrointestinal abnormalities – especially exomphalos.
- Renal abnormalities.
- Severe learning difficulties.

Patau's syndrome (trisomy 13)

The prevalence is 1 in 6000 births and the majority die in the first few days or weeks of life. Maternal age is less of a predisposing factor than for trisomies 18 and 21 and the recurrence risk is low.

Clinical features

- IUGR.
- Holoprosencephaly. Developmental defect of the forebrain resulting in a single cerebral hemisphere. Cyclops in the most severe form.
- Craniofacial abnormalities. Microcephaly, hypotelorism, small nose, cleft lip and palate (80%), microphthalmos.
- Scalp defects.
- Polydactyly.
- Renal and cardiac abnormalities.

Turner's syndrome (XO)

This is the only monosomy that is compatible with life. The prevalence is 0.4 per 1000 live births and it is a common finding in first-trimester abortuses. The lifespan may be normal but sterility is almost invariable. Fifty-five per cent are 45XO, but the rest are mosaics or have other abnormalities.

Clinical features

- Cystic hygroma and hydrops in second trimester.

- Lymphoedema at birth, especially hands and feet and back of neck (nuchal) swelling which can be detected on antenatal ultrasound.
- Short stature due in part to skeletal dysplasia.
- Webbed neck, low posterior hairline, widely spaced nipples, nail changes, cubitus valgus (increased elbow carrying angle), tendency to keloid scarring.
- Congenital heart disease, particularly coarctation of aorta (hypertension).
- Renal tract abnormalities. Common but not usually clinically significant, e.g. duplex ureters, horseshoe kidney.
- Streak ovaries, primary amenorrhoea and failure of development of secondary sexual characteristics. Puberty is induced with oestrogen replacement but the patient remains infertile. Osteoporosis in adult life occurs in the absence of oestrogen replacement.
- Intelligence. Normal or mild learning difficulties.

Fragile X syndrome

This is an inherited X-linked disorder which was first described in 1969. It is the commonest cause of learning difficulties in males after Down's syndrome, with a prevalence of 1 in 1000 males. Analysis of chromosomes with special culture techniques identifies a fragile site on the long arm of the X chromosome (Xq27 folate-sensitive fragile site) in a proportion of cells (5 – 20%). Fragile sites may be seen in female carriers but the absence of sites does not exclude carrier status.

Clinical features

- Learning difficulties. Moderate to severe in males, mild to moderate in a third of female carriers.
- Craniofacial features. Large head (OFC often above 97th centile), large jaw and ears, prominent forehead.
- Large testes. Particularly after puberty.
- Connective tissue dysfunction. May present like Ehlers–Danlos or Marfan's syndromes. Mitral valve prolapse is not uncommon.

Further reading

Brook CGD. The short child. In: Brook CGD, ed. *A Guide to the Practice of Paediatric Endocrinology.* Cambridge University Press, 1993: 42–5.

Smith DW. *Recognizable Patterns of Human Malformation*, 4th edn. Philadelphia: WB Saunders, 1988.

Winter RM. Fragile X mental retardation. *Archives of Disease in Childhood,* 1989; **64:** 1223–4.

Related topics of interest

Antenatal diagnostics (p. 23)
Growth – short and tall stature (p. 149)

CHRONIC DIARRHOEA

Persistent diarrhoea presents to the health professional when the frequency or looseness of the child's stools is perceived by the parent to be excessive. However, there is a wide variation in the 'normal' pattern, depending on the age and diet of the individual child. Constipation with 'overflow', i.e. semiliquid stool seeping past an impacted faecal mass, often with incontinence, may present as chronic diarrhoea, and this diagnosis should be considered at an early stage.

Toddler diarrhoea (chronic non-specific diarrhoea of infancy)

Typically, a pattern of variable diarrhoea occurs in a child of 6–24 months who is otherwise well and thriving. Constipation and diarrhoea may alternate, and both mucus and undigested food matter are often seen in the stool. Diagnosis depends on the demonstration of normal growth and a clinical exclusion of other pathology. However, sucrose intolerance and infestation with *Giardia* may cause chronic diarrhoea with normal growth, and these conditions may need to be excluded by appropriate investigation (see below). Parents should be reassured that the problem is likely to resolve spontaneously, and a simple explanation of intestinal motility is often helpful. Exclusion of cow's milk and egg under dietetic supervision may help in those children with a history of atopy, whereas other children may benefit from a diet high in polyunsaturated fat.

Protracted diarrhoea

This is defined as the passage of four or more loose stools a day for over 2 weeks, with associated failure to thrive. Occasionally at presentation, the child may be severely ill, and resuscitation must then precede thorough investigation. A wide range of conditions must then be considered.

Aetiology of protracted diarrhoea

1. *Infection. Giardia lamblia,* enteropathogenic *Escherichia coli*, hookworm.

2. *Inflammation.* Ulcerative colitis, Crohn's disease.

3. *Protein intolerance.* Coeliac disease, cow's milk or soy protein intolerance.

4. *Carbohydrate intolerance.* Lactase deficiency, secondary lactose intolerance.

5. *Surgical conditions.* Hirschsprung's disease, malrotation, blind loop syndrome.

6. Immunodeficiency. AIDS, severe combined immunodeficiency.

7. Pancreatic insufficiency. Cystic fibrosis, Schwachman syndrome.

8. Other disorders. Congenital chloridorrhoea, lymphangiectasia, small bowel lymphoma, abetalipoproteinaemia, autoimmune enteropathy, microvillus atrophy, Munchausen by proxy syndrome.

Clinical assessment

The history must establish the exact timing of the onset of diarrhoea, e.g. from birth, after recent acute gastroenteritis or following travel abroad. Introduction of new food substances such as gluten or cow's milk may also be relevant. Systematic enquiry should seek to exclude features of generalized chronic disease such as cystic fibrosis. Family history may be important in coeliac disease (approximately 10% of siblings are at risk), the inborn errors of metabolism and cystic fibrosis. A clear account of the stool frequency, consistency and presence of blood or mucus must be obtained, and any associated features (abdominal pain, allergic manifestations) should be noted. Examination must include assessment of nutritional status (weight, length, head circumference, skinfold thickness and mid-arm circumference) and dehydration. There may be specific signs of disease, e.g. respiratory signs in cystic fibrosis, but often no diagnostic clues are present.

Investigation

Certain selected investigations only may be required, depending on clinical assessment.

- FBC, with film, ESR. Coagulation studies, especially prior to jejunal biopsy.
- U&E, creatinine and acid–base balance.
- Serum calcium, phosphate, alkaline phosphate, ferritin and folate. Trace metals (zinc, copper).
- Antigliadin antibodies, endomysium antibodies.
- Stool specimens for microbiology (MC&S, virology, parasites), electrolytes, reducing substances, chromatography, fat globules.
- Sweat test.
- Radiological investigations: abdominal radiograph, barium meal and follow-through.
- Immunological profile, including functional assessment.

- Jejunal biopsy.
- Pancreatic function tests, e.g. stool chymotrypsin, pancreolauryl test, aspiration of duodenal juice.
- Other tests: colonoscopy and biopsy, breath hydrogen test, thyroid function tests, congenital infection screen, gastrointestinal hormone profile, gut auto-antibody screen.

The management approach will be determined by the degree of illness of the child and the facilities available locally. It may be appropriate in certain cases to refer the child to a centre where specialist diagnostic and supportive services are available.

Further reading

Devane SP, Candy DCA. Investigation of chronic diarrhoea. *Current Paediatrics,* 1992; **2(4):** 189–93.

Puntis JWL. Assessment of pancreatic exocrine function. *Archives of Disease in Childhood,* 1993; **69:** 99–101.

Related topics of interest

Coeliac disease (p. 82)
Constipation (p. 101)
Cystic fibrosis (p. 104)
Failure to thrive (p. 133)
Gastroenteritis (p. 138)

CHRONIC RENAL FAILURE

Renal insufficiency results from loss of functioning nephrons, but there is remarkable renal reserve. The glomerular filtration rate (GFR) can fall by over 50% before nitrogenous waste products (urea and creatinine) accumulate and is reduced to 25% before symptoms and signs of renal failure occur. The incidence of chronic renal failure in the UK is 2–3 childen per million population per year.

Problems

- Anaemia.
- Hypertension.
- Hyperkalaemia.
- Growth failure.
- Renal osteodystrophy.
- Hyperuricaemia.
- Type IV hyperlipidaemia.

Aetiology

- Glomerulonephritis (30–35%).
- Reflux nephropathy / chronic pyelonephritis (~ 20%).
- Renal tract malformations (25–30%) – hypoplastic, dysplastic or cystic kidneys, obstruction, e.g. posterior urethral valves.
- Hereditary renal disorders (~10%), e.g. Alport's syndrome, cystinosis.
- Haemolytic uraemic syndrome (1–2%).
- Cortical or tubular necrosis (1–2%).

Clinical features

The onset of symptoms is usually insidious, but acute deterioration may be precipitated by intercurrent infection or dehydration.

1. GFR 20–25% of normal (chronic renal failure)

- Polyuria, polydipsia, enuresis, failure to thrive, hypertension.

2. GFR 5–20%

- Anorexia, metabolic acidosis, failure to thrive, bone pain (renal osteodystrophy).

3. GFR < 5% (end-stage renal failure)

- Oliguria and oedema, anaemia, lethargy, itching, nausea. Dialysis or transplantation necessary for survival.

Investigation

- *U&E creatinine, bicarbonate*. Raised urea and creatinine, hyperkalaemia, metabolic acidosis.
- *Glomerular filtration rate*. Measured by either creatinine clearance or ^{51}Cr-EDTA clearance.

- *Serum calcium, phosphate, alkaline phosphatase.* Hyperphosphataemia, hypocalcaemia, raised alkaline phosphatase. Hyperphosphataemia results from impaired renal excretion and leads to decreased absorption of calcium. Hyperphosphataemia also causes a secondary increase in parathyroid hormone (PTH) secretion, which promotes calcium resorption (renal osteodystrophy). Bone resorption is increased by acidosis. Relative vitamin D resistance occurs as a result of impaired renal hydroxylation of 25-OH calciferol to 1,25-OH calciferol, resulting in reduced calcium absorption from the gut. Although the total serum calcium is low, symptoms are rare since acidosis increases the proportion of ionized calcium.
- *Full blood count.* Normochromic normocytic anaemia due to decreased red cell production and shortened red cell lifespan. Iron and folic acid deficiency may result from poor dietary intake. Erythropoietin levels are low.
- *Uric acid.* Raised and may worsen renal failure by causing a uric acid nephropathy.
- *Lipids.* Raised triglycerides and cholesterol.
- *Radiology.* Renal tract ultrasound, intravenous urography (IVU), MCU and isotope scans, may help identify cause. Radiographs of wrists, hands and knees to identify renal osteodystrophy.
- *Renal biopsy.*

Management

Chronic renal failure is rarely reversible so management is aimed at preserving the remaining renal function and minimizing metabolic disturbances. These children are best managed by an established multidisciplinary team within a specialist renal unit.

1. Hypertension. This is due to the primary renal disease or to salt and water overload. If treated promptly and effectively the rate of progression of renal failure may be slowed and there may even be a modest rise in GFR. Beta-blockers and diuretics are used, with a vasodilator such as hydralazine if severe. Acute hypertensive crises may occur and require urgent intravenous therapy.

2. Renal osteodystrophy. The clinical consequences of renal failure on bone and calcium metabolism are:

- Osteitis fibrosa cystica (due to secondary hyperparathyroidism).

- Rickets (due to vitamin D resistance).
- Ectopic calcification (due to an increased calcium/phosphate product despite hypocalcaemia).

Early recognition is important, and serum calcium, phosphate, and alkaline phosphatase should be monitored in addition to radiology. A rise in PTH may be the best early indicator. Specific treatments include dietary restriction of phosphate, oral phosphate-binding agents (e.g. calcium carbonate) and 1,25 dihydroxy vitamin D (calcitriol) or 1α-OH calciferol (a vitamin D analogue) supplements.

3. Anaemia. Iron and folic acid supplements are given (serum ferritin should be monitored to prevent iron overload). Erythropoietin can be given intravenously or subcutaneously to maintain the haemoglobin. Transfusions may aggravate hypertension and fluid overload.

4. Nutrition. An adequate calorific intake is essential to maintain growth and good nutritional status. Dietary supplements and nasogastric feeding should be considered early. The degree of protein restriction depends on the GFR, and fluid intake depends on the urine output and state of hydration. The calcium intake should be adequate and phosphate intake restricted. Vitamin supplements are given in the form of Ketovite liquid and tablets. Oral bicarbonate supplements may be required if acidosis is marked.

5. Hyperkalaemia. Severe or worsening hyperkalaemia is an indication for dialysis. Calcium resonium or nebulized salbutamol is used to reduce serum potassium levels acutely.

6. Dialysis. At least five children per million population are accepted on to dialysis programmes in Europe each year. Indications for dialysis vary, but it should be considered when fluid overload, electrolyte disturbances, acidosis and symptomatic uraemia remain uncontrolled despite conservative treatment. Continuous ambulatory peritoneal dialysis (CAPD) is the preferred method, and only a small minority receive haemodialysis. Complications of CAPD include peritoneal catheter site infections and peritonitis.

7. Renal transplantation. Renal transplantation is now the established treatment for end-stage renal failure in children. In the UK more than 80% of first cadaver donor kidneys are functioning at 1 year and there is an almost 100% success rate with live related donor kidneys.

Further reading

Mehls O, Fine RN. Chronic renal failure in children. In: Cameron S, Davison AM, Grünfeld JP, Kerr D, Ritz E, eds. *Oxford Textbook of Clinical Nephrology*. Oxford: Oxford University Press, 1992; 1605–21.

Trompeter RS. Renal transplantation. *Archives of Disease in Childhood,* 1990; **65:** 143–6.

Related topics of interest

Calcium metabolism (p. 47)
Glomerulonephritis (p. 146)
Urinary tract infection (p. 322)

CLEFT LIP AND PALATE

Cleft lip and palate occurs in 1 in 700 births and is usually an isolated abnormality, but is occasionally part of a chromosomal syndrome or sequence of malformations. There is often a family history, and after one affected child the recurrence risk is 1 in 25. If one parent is affected the risk of an affected child is 1 in 20. Most cases are determined by polygenic inheritance but occasional cases are linked to environmental factors or maternal medication. Affected children encounter both short- and long-term problems which are best managed by a multidisciplinary team.

Problems

- Feeding difficulties.
- Maternal bonding.
- Respiratory problems.
- Speech delay.
- Serous otitis media (glue ear).
- Dental problems.
- Cosmetic appearance.

Aetiology

- Idiopathic.
- Polygenic inheritence; often a positive family history.
- Maternal drugs, e.g. corticosteroids, phenytoin.
- Environmental factors are ill defined.
- Chromosomal abnormalities, e.g. trisomy 13 (Patau's syndrome).
- Pierre Robin sequence. Cleft palate occurs secondary to severe micrognathia.

Normal migration and fusion of mesodermal folds forming the lip, alveolus and anterior hard palate occurs between the fifth and seventh weeks of gestation. Failure of fusion results in unilateral or bilateral clefts. The remainder of the palate fuses in the midline between the ninth and 12th weeks. Tissue migration may be disorganized or obstructed by the tongue, resulting in midline cleft.

Clinical features

One-third have a cleft lip only. This is more common in boys (3M:2F), and is more common on the left than the right. One-quarter have a cleft palate only, and this is more common in girls (3F:2M). The remainder have both cleft lip and palate.

Clefts range in severity from a notch in the lip to complete separation of the maxillae. A bifid uvula may indicate the presence of a submucous cleft (poor muscle union in the soft palate), which can lead to difficulties with speech. Clefts are associated with abnormalities of the cardiovascular, renal, central nervous and gastrointestinal systems, particularly when part of a chromosomal syndrome. Respiratory embarrassment occurs in the Pierre Robin sequence owing to posterior displacement of the tongue secondary to underdevelopment of the mandible. Nursing prone or insertion of a nasopharyngeal airway usually avoids the need for tracheostomy. The respiratory embarrassment resolves spontaneously as the mandible grows, these children usually having a normal profile by 4–6 years of age.

Management

1. Short-term problems

- *Feeding.* Breast-feeding can be successful in some cases. Various special teats are available for bottle-feeding. Orthodontic plates may be helpful.
- *Maternal bonding.* Before and after surgery photographs of a similar cleft may be helpful.

2. Surgery. The precise timing of surgery varies from centre to centre. The lip is closed early (neonatal period to 3 months) and the palate is closed later (6–15 months). Preoperative orthodontic splints are sometimes used to align the aleolar margins and to reduce the size of the cleft.

3. Longer term problems. These are best managed by a multidisciplinary team. The paediatrician ensures coordination of the team, and monitors growth and development. Surgery is usually undertaken by plastic or maxillofacial surgeons. As the child grows there may need to be revisions to achieve the best cosmetic result. Orthodontic review is important to ensure dental alignment. Audiological assessment is recommended as Eustachian tube obstruction and subsequent serous otitis media is common. Grommets may be needed. Speech delay due to mechanical problems with the palate, exacerbated by glue ear, mean that a speech therapist is a vital member of the team.

Further reading

Davies D. Cleft lip and palate. *BMJ*, 1985; **290:** 625–8.

Holder SE. Cleft lip – is there light at the end of the tunnel? *Archives of Disease in Childhood,* 1991; **66:** 829–32.

Watson JD, Pigott RW. Management of cleft lip and palate. *Hospital Update*, 1991; April: 306–12.

Related topic of interest

Hearing and speech (p. 165)

COELIAC DISEASE

The basic aetiology of coeliac disease is unknown. Malabsorption is accompanied by characteristic morphological abnormalities in the small intestinal mucosa, which revert to normal when gluten is removed from the diet. Relapse occurs when gluten-containing foods are reintroduced into the diet. The prevalence in the UK is fairly constant at 1 in 2000, with a much higher incidence in Ireland and a much lower incidence outside Europe. Ten per cent of patients have an affected first-degree relative, and there is a high prevalence (> 80%) of HLA groups B8, DR3, and DQw2 and also of DR7 and DR2G in affected patients.

Problems

- Malabsorption.
- Abdominal pain and distension.
- Growth failure.
- Dermatitis herpetiformis.
- Malignancy.

Pathology

The proximal small bowel mucosa is predominantly affected, and it is the α-gliadin peptide fraction of the gluten in wheat, rye, barley and oats which causes damage. The lesion is described as subtotal villous atrophy, with loss of the villous pattern. There is compensatory crypt hypertrophy, and the surface columnar epithelial cells become cuboidal, with increased intraepithelial lymphocytes. There are increased numbers of plasma cells and lymphocytes in the lamina propria.

Clinical features

At a variable period after introduction of cereals into the diet, failure to thrive may become a feature, accompanied in many cases by abdominal symptoms. Wasting is sometimes marked, e.g. over the buttocks. Stools may be loose, pale, bulky and offensive. Abdominal distension, anorexia, vomiting and abdominal pain are common symptoms. Typically, the child is miserable and irritable. Iron deficiency anaemia with hypochromic indices is a frequent finding and may be the major presenting feature, although folate levels may also be low. Hypoalbuminaemia with oedema, hypoprothrombinaemia and vitamin D deficiency are rarer findings.

Investigation

- FBC, red cell folate, ferritin.
- Coagulation screen – abnormalities should be corrected before jejunal biopsy is performed.
- Serum albumin, calcium, phosphate, alkaline phosphatase.
- Antigliadin, antiendomysium and antireticulin antibodies.

- Jejunal biopsy.

Other screening tests such as the xylose absorption test and barium follow-through studies may also be abnormal, but are not routinely performed.

Diagnostic criteria

The European Society of Paediatric Gastroenterology and Nutrition (ESPGAN) revised and clarified the practical approach to diagnosis of coeliac disease in 1989. The initial step in diagnosis remains the finding of characteristic small bowel mucosal changes on histological examination of a biopsy specimen. This should be followed by an unequivocal clinical remission on a strict gluten-free diet. A gluten challenge is not mandatory in all patients, but is recommended if there was doubt about the initial diagnosis, e.g. no initial biopsy performed, or doubtful histology. Also, in children under 2 years old, there is a high incidence of other causes of enteropathy, e.g. post-infectious and cows' milk-sensitive enteropathy, and a later gluten challenge is advisable. When gluten challenge is performed, a control biopsy taken when the patient is on a gluten-free diet is essential, taking a further specimen 3–6 months after challenge (or earlier if clinical relapse), and again at 2 years after challenge if the patient remains well. Very late relapses (up to 7 years) have been reported, and prolonged clinical monitoring is needed. The presence of circulating antibodies to gliadin, reticulin, endomysium and jejunum is a valuable adjunct to the clinical assessment, and is of value for preliminary screening, in monitoring clinical response and in detecting relapse and non-compliance.

Management

The clinical response to a gluten-free diet is usually seen within a few weeks, with an improvement in temperament, abdominal symptoms and weight gain. Dermatitis herpetiformis responds to the diet, but also to dapsone. Replacement haematinics may be needed initially. Expert dietary advice should be given, and parents should make contact with the Coeliac Society for further practical help. Adolescents may question the need to continue their diet, and this is a further indication to consider a challenge biopsy, since once the diagnosis is firmly established the diet should be life-long. Dietary compliance may possibly be protective against coeliac-associated malignancies (T-cell lymphoma and carcinoma of jejunum and oesophagus) in adult life.

Further reading

Report of the Working Group of ESPGAN. Revised criteria for diagnosis of coeliac disease. *Archives of Disease in Childhood,* 1990; **65:** 909–11.
Trier JS. Celiac sprue. *New England Journal of Medicine*, 1991; **325:** 1709–18.

Related topics of interest

Chronic diarrhoea (p. 72)
Failure to thrive (p. 133)

COMA

The major causes of altered consciousness level in childhood are head injury, CNS infections, hypoxic–ischaemic events and metabolic disturbances. Acute encephalopathy is manifested by a combination of altered consciousness level and seizures, and is a major cause of death and neurological sequelae. Primary management of coma is aimed at cardiopulmonary resuscitation and support. Secondary management aims to determine and treat the cause.

Aetiology

- Infection (35%). Meningitis, encephalitis.
- Cerebral anoxia and ischaemia. Shock, cardiorespiratory arrest, drowning.
- Trauma. Accidental, non-accidental.
- Drugs and poisons. Alcohol, barbiturates, lead, Munchausen by proxy.
- Metabolic.
  - Diabetes mellitus (hypoglycaemia, ketoacidosis).
  - Hepatic failure, Reye's syndrome.
  - Uraemia (haemolytic uraemic syndrome).
  - Hyponatraemia.
  - Inborn errors of metabolism.
  - Post-burns encephalopathy.
- Vascular. Intracranial bleeding (e.g. arteriovenous malformation), thrombosis (e.g. DIC, sagittal sinus).
- Acute hypertension.
- Epilepsy. Post-ictal, status epilepticus.
- Intracranial tumour.
- Haemorrhagic shock encephalopathy.

Primary assessment and management

1. Airway. Clear and maintain.

2. Breathing. Give oxygen. If the patient is hypoventilating or apnoeic, intubate and ventilate.

3. Circulation. Insert an i.v. line, take bloods for investigations (including BM Stix). Check blood pressure and pulse rate. If the patient is hypotensive give a 20 ml/kg plasma bolus over 5–10 minutes. If the patient is hypoglycaemic, give an i.v bolus of 25% dextrose and then commence a 10% dextrose infusion.

4. Obtain a brief history of the illness/accident from an accompanying adult (parent, paramedic) while initial assessment and management are proceeding, including any relevant past medical history/drugs.

5. Assessment of neurological status. Assess consciousness level using the AVPU system (alert, responds to voice, responds to pain, or unresponsive). Assess coma score prior to sedative or paralysing drugs, using the Glasgow coma scale (best motor, verbal and eye-opening responses – a modified scale is available for children under 4 years). Note any decorticate (flexed arms, extended legs) or decerebrate (extended arms and legs) posturing. Note pupillary dilation, inequality or unresponsiveness (third nerve compression) which suggests tentorial herniation of the ipsilateral hippocampal gyrus. Small pupils are seen with narcotic, barbiturate and phenothiazine poisoning. Bilaterally dilated pupils are seen post-ictally, in late barbiturate poisoning and with irreversible brain damage. Tentorial herniation may also cause a sixth nerve palsy. Hypertension, bradycardia and an irregular respiratory pattern are signs of cerebellar tonsil herniation through the foramen magnum ('coning').

6. Clinical examination. Note any evidence of sepsis (rash, neck stiffness) or trauma (including CSF leakage from ears or nose). The breath may smell of alcohol. Hepatomegaly is a feature of Reye's syndrome and some inborn errors of metabolism.

7. Treat any identifiable causes, e.g. DKA, and treat any seizures. If sepsis is suspected treat with broad-spectrum antibiotics.

Initial investigation

- FBC and clotting studies.
- U&E, creatinine, calcium.
- Glucose.
- Blood culture.
- Arterial blood gas.
- Group and save serum (cross-match if the patient has sustained major trauma).

Secondary assessment and management

The cause of coma may be apparent from the primary assessment, but if it is not then a thorough history and examination with further investigations should follow. Further neurological assessment should include oculocephalic (doll's eye) reflexes (avoid in children with neck injuries), fundoscopy (papilloedema may not occur if there is an acute rise in ICP, retinal haemorrhages may indicate non-accidental injury) and examination for lateralizing signs.

Whatever the cause of coma, it is vital to maintain homeostasis.

- Ensure good oxygenation.
- Maintain blood pressure.
- Normalize acid–base and electrolyte balance.
- Maintain blood glucose.
- Maintain body temperature.
- Detect and treat raised ICP.

Further investigation

- Urine culture.
- Urinalysis. For glucose, protein, blood.
- Blood and urine for toxicology.
- Liver function tests and ammonia.
- Lactate, pyruvate.
- Radiology (CT scan, US, chest radiographs).
- Lumbar puncture. Contraindicated until evidence of raised intracranial pressure is excluded by CT scan. CSF for bacteriology, virology, metabolic investigations.

Raised intracranial pressure (ICP)

Most causes of coma are accompanied by a degree of raised ICP, resulting from either an expanding space-occupying lesion (e.g. haematoma) or cerebral oedema. The aims of treatment are to maintain adequate cerebral blood flow for the brain's metabolic needs and to prevent the ICP rising to a level which will result in brain herniation. Cerebral blood flow is dependent on both cerebral perfusion pressure (CPP = mean arterial blood pressure minus ICP) and cerebrovascular resistance (controlled by vascular autoregulation, which may be impaired by trauma, hypoxia or ischaemia).

The child should be nursed head up with the head in the midline to encourage cerebral venous drainage. A raised arterial $P\text{CO}_2$ will cause cerebral vessels to dilate, leading to a further rise in ICP. The role of hyperventilation to maintain a low $P\text{CO}_2$ is controversial since it may reduce an already critical cerebral blood flow, but a low–normal level is recommended. Mannitol reduces cerebral oedema by raising the plasma osmolality and drawing water into the intravascular space, but dangers include hyperviscosity and prerenal failure. Barbiturates (e.g. thiopentone) reduce the cerebral metabolic rate but may cause systemic hypotension. Hypothermia will reduce metabolic requirements but may be impractical. There is no good evidence that fluid restriction reduces ICP, although it is vital that hyposmolar intravenous fluids are avoided. The role of invasive ICP monitoring remains controversial, as the exact level at which brain dysfunction occurs is unclear, but it is useful to monitor the effects of treatments such as mannitol.

Further reading

Coma. In: Advanced Life Support Group. *Advanced Paediatric Life Support – the Practical Approach.* London: BMJ Publishing, 1993: 111–20.

Sharples PM. Intracranial pressure monitoring in comatose children: concepts and controversies. *Current Paediatrics.* 1991; **1:** 46–8.

Related topics of interest

Hypertension (p. 177)
Inborn errors of metabolism (p. 191)
Meningitis and encephalitis (p. 220)
Shock (p. 291)

CONGENITAL DISLOCATION OF THE HIP AND TALIPES

Problems

- Mobility.
- Association between talipes and congenital dislocation of the hip (CDH).
- Association with neuromuscular disorders (hypotonia, spina bifida).
- Late diagnosis of CDH.

Congenital dislocation of the hip (CDH)

CDH occurs in 1.5 per 1000 live births and is six times more common in girls than in boys. There is an increased incidence in first-degree relatives of an affected individual. Dislocation occurs secondary to abnormal laxity of the joint capsule with elongation of the ligamentum teres. There may be a shallow acetabulum and hypoplasia of the femoral head. Unilateral defects are twice as common as bilateral defects. Predisposing factors include breech delivery, spina bifida and hypotonia. It is essential to detect CDH early as the treatment is relatively simple. Late signs include asymmetry of the thigh folds, shortening and external rotation of the limb and a waddling gait with a positive Trendelenburg test.

Investigation

1. Routine physical examination. Ortolani's and Barlow's tests are used in the newborn period and throughout early infancy to detect dislocated or dislocatable hips. With the hips and knees flexed, the thighs are abducted and a dislocated femoral head will 'clunk' back into the acetabulum. Application of backward pressure as abduction is commenced will detect a dislocatable hip. By 2–3 months of age instability of the joint is lost and important signs are asymmetrical skin folds and limitation of abduction.

2. Radiology is not useful in early detection of CDH as interpretation of radiographs of the unossified hip is difficult.

3. Ultrasound is used in some centres as a screening procedure, particularly for those at higher risk of CDH, e.g. family history, breech delivery.

Management

When an unstable or dislocated hip is detected early, treatment should be commenced immediately with an abduction splint (e.g. Von Rosen). This maintains the femoral head in the correct position until the acetabulum has developed sufficiently to prevent dislocation (usually 2–4 months). The splint must not be removed during this time and can cause parents difficulties, particularly with nappy changing and bathing. Occasionally dislocated hips cannot be reduced in a splint and these require surgical intervention. CDH which is detected late is difficult to treat because of the relatively fixed displacement of the femoral head with limitation of abduction and adductor contracture. Surgical correction with a period of traction may be successful.

Talipes

Talipes occurs in 1 in 1000 live births and is three times more common in boys than girls. In 50% of cases the disorder is bilateral, and predisposing factors include oligohydramnios, spina bifida and a family history of talipes. The most common deformity consists of fixed plantar flexion with inversion of the foot and adduction of the forefoot (talipes equinovarus). A less common deformity consists of dorsiflexion and eversion of the foot (talipes calcaneovalgus). All cases must be carefully examined for associated hip and spine problems.

Management

Gradual manipulative correction with adhesive strapping over a period of 6–8 weeks will correct the majority of the less rigid deformities. Surgical correction with soft-tissue release and tendon transplantation to correct the inversion of the hindfoot, followed by splinting with Denis Browne boots, may be necessary for more rigid deformities. Corrective bone surgery is sometimes necessary in later childhood.

Further reading

Hensinger RN, Jones ET. Orthopaedic problems in the newborn. In: Roberton NRC, ed. *Textbook of Neonatology*, 2nd edn. Edinburgh: Churchill Livingstone, 1992; 889–914.

Related topic of interest

Limping and joint disorders (p. 205)

CONGENITAL HEART DISEASE – ACYANOTIC

Congenital heart disease (CHD) is the commonest serious congenital abnormality, occurring in 8 per 1000 live births. In 10–15% of affected children there is an associated non-cardiac anomaly. Acyanotic lesions account for the majority of defects (more than 70%) and early death is uncommon. The risk of a subsequent child having CHD is increased three times.

Problems

- Heart failure.
- Failure to thrive.
- Arrhythmias.
- Subacute bacterial endocarditis (SBE).
- Hypertension.
- Sudden death.
- Pulmonary hypertension.

Aetiology

- Idiopathic / multifactorial inheritance.
- Chromosomal abnormalities
  50% of children with Down's syndrome have cardiac defects (particularly septal defects).
  10% of children with Turner's syndrome have coarctation.
  Cardiac defects are common in children with trisomy 18 (Edward's syndrome) and trisomy 13 (Patau's syndrome).
- Teratogens, e.g. congenital rubella, alcohol.
- Maternal conditions e.g. diabetes.
- Single gene defects, e.g. Marfan's and Noonan's syndromes.
- Associated conditions, e.g. de Lange syndrome (venticular septal defect).

Presentation

1. *Asymptomatic murmur.* An innocent murmur is soft, short, systolic and varies with the position of the child. Features such as a loud, long or diastolic murmur, a thrill or abnormal heart sounds suggest a significant murmur.

2. *Heart failure.* Tachycardia, tachypnoea, hepatomegaly and sweating particularly on feeding usually develop over the first 6 weeks of life as pulmonary vascular resistance falls and left-to-right shunting increases.

3. *Hypertension.* Hypertension may be the only sign of coarctation of the aorta in older children.

Ventricular septal defect (VSD)

This is the commonest heart defect and it usually presents with a murmur or heart failure. The murmur is pansystolic, loudest at the lower left sternal edge, and often associated with a thrill. In 50% of cases the defect closes spontaneously. Progressive shortening of the murmur and an increasingly loud pulmonary component of the second heart sound are evidence of pulmonary hypertension, which may develop if the left-to-right shunt is large. Eventual reversal of the shunt leads to cyanosis (Eisenmenger's syndrome).

1. Chest radiography. Normal, or cardiomegaly and pulmonary plethora.

2. ECG. Normal or biventricular hypertrophy. Right venticular hypertrophy (RVH) increases as pulmonary hypertension develops.

Atrial septal defect (ASD)

An ASD usually presents with an asymptomatic ejection systolic murmur in the pulmonary area with fixed splitting of the second heart sound. This murmur is due to increased flow over the pulmonary valve and there may be a diastolic murmur in the tricuspid area. Spontaneous closure is very unusual. Pulmonary hypertension or heart failure may develop in adult life. SBE is very rare. The majority are secundum defects due to failure of closure of the foramen ovale. Primum defects are due to failure of development of the septum primum.

1. Chest radiography. Normal or cardiomegaly with prominent right atrium and pulmonary arteries.

2. ECG
- Left axis deviation in primum defects.
- Normal or right axis deviation in secundum defects.
- Incomplete right bundle branch block.

Pulmonary stenosis (PS)

This presents with an asymptomatic ejection systolic murmur in the pulmonary area, radiating to the back and often associated with a thrill. The second heart sound is normal.

1. Chest radiography. Normal or post-stenotic dilatation of the pulmonary artery.

2. ECG. RVH increases with severity.

Aortic stenosis (AS)

AS often presents with an asymptomatic murmur, but if severe can cause heart failure or sudden death. There is an ejection systolic murmur in the aortic area radiating to the carotids. An ejection click may be heard.

1. Chest radiography. Normal or post-stenotic dilatation of the aorta.

2. ECG. Left ventricular hypertrophy (LVH) and strain increases with severity.

Patent ductus arteriosus (PDA)

This is classically described as causing a continuous machinery murmur under the left clavicle, which extends through systole and the second heart sound into diastole. The peripheral pulses are full and collapsing, with easily palpable foot pulses. It is common in premature infants who may present with heart failure. In older children it is usually asymptomatic.

Both chest radiographs and the ECG are normal unless there is a very large left-to-right shunt.

Coarctation of the aorta

Coarctation presents with hypertension in the arms, with or without heart failure. Absent femoral pulses or radiofemoral delay may be the only signs, or a systolic murmur may be heard. The most severe form leads to an interrupted aortic arch, which presents with heart failure in the neonatal period when the ductus arteriosus closes.

1. Chest radiography may show cardiomegaly but is often normal. Rib notching due to collaterals is sometimes seen in older children.

2. ECG. LVH increasing with severity.

Management of acyanotic CHD

1. Medical. Management includes the treatment of heart failure and hypertension. Adequate nutrition is important, particularly if the child is in heart failure as calorie requirement will be increased. Parents should be advised about antibiotic prophylaxis against SBE, and about the importance of immunizations. Most children with CHD will limit their own exercise and the only condition in which strenuous exercise, should be avoided is moderate to severe aortic stenosis because of the risk of sudden death.

2. Surgical. Septal defects which will not close spontaneously are closed surgically. Reduction of pulmonary blood flow by pulmonary artery banding is sometimes necessary with large defects to reduce symptoms of heart failure and to prevent the development of pulmonary hypertension. The defect is then repaired at a second operation months to several years later, when the child has grown.

If outflow obstruction (e.g. AS, PS, coarctation) is marked with increasing ventricular hypertrophy or heart failure, repair is indicated. Balloon angioplasty is becoming increasingly popular, particularly for coarctation, but the obstruction often recurs.

Further reading

Anderson RH, Macartney FJ, Shinebourne EA, Tynan M. (eds) *Paediatric Cardiology.* Edinburgh: Churchill Livingstone, 1987.

Related topics of interest

Congenital heart disease – cyanotic (p. 95)
Hypertension (p. 177)

CONGENITAL HEART DISEASE – CYANOTIC

Cyanotic lesions account for only about 25% of congenital heart disease (CHD), but many are life-threatening in the first few weeks of life. Even with surgery, 15–20% of infants will die in the first year of life. Cyanosis is clinically detectable when reduced haemoglobin exceeds 5 g/100 ml and is more easily recognized in the newborn with a high haemoglobin than in an anaemic child. Cyanosis may occur because the lungs are underperfused owing to a right-to-left shunt bypassing the lungs (e.g. Fallot's tetralogy) or because there is mixing of the systemic and pulmonary circulations (e.g. transposition of the great arteries with VSD). In the latter case, lung perfusion is normal or excessive. Cyanosis may be apparent in the neonatal period or become increasingly obvious over the first few months of life, particularly when the baby feeds or cries. In the older child, cyanosis is usually obvious at rest, exercise tolerance is reduced and clubbing occurs.

Problems

- Hypoxia.
- Acidosis.
- Heart failure.
- Failure to thrive.
- Polycythaemia.
- Thromboembolic events.
- Hypercyanotic spells.

The blue baby

Cyanosis in the first week of life may be due to respiratory disease, neurological depression (asphyxia, drugs), persistent fetal circulation or cyanotic CHD. The baby with CHD usually has no respiratory difficulty unless there is marked metabolic acidosis or heart failure. There may be a murmur and the second heart sound is usually single. If the atrioventricular septum is intact, the baby will become acutely cyanosed and collapse as the ductus arteriosus begins to close. The most important lesions in descending order of frequency are transposition of the great arteries (TGA), pulmonary atresia (PA) and tricuspid atresia (TA).

Investigation

- Arterial blood gases. Low $P\text{O}_2$ and a normal $P\text{CO}_2$, with or without metabolic acidosis.
- CXR
  TGA: plethoric lung fields, egg shaped heart + narrow vascular pedicle.
  PA/TA: oligaemic lung fields.
- ECG
  TGA: right axis deviation, RVH
  PA: normal/left axis, LVH since pulmonary circulation maintained by left ventricle via patent ductus.

TA: marked left axis deviation leading to a superior axis, LVH, RAH.

- Hyperoxia (nitrogen washout) test. To differentiate between heart and lung disease as the cause of cyanosis. $Po_2 > 14$ kPa (100 mmHg) in 80–100% oxygen excludes a major right-to-left shunt.
- Echocardiography and cardiac catheterization.

Management

If there is no mixing of oxygenated and deoxygenated blood other than via the ductus arteriosus, prostaglandin E_2 is used to keep the ductus open until a septostomy or surgery can be performed. A balloon atrial septostomy (Rashkind's procedure) can be performed during cardiac catheterization. Initial surgery aims either to increase pulmonary blood flow by a systemic to pulmonary shunt (e.g. Blalock–Taussig shunt) or to correct the defect anatomically (e.g. arterial switch for TGA). Later surgery anatomically corrects the defect or redirects blood through the correct circulations (e.g. Mustard procedure for TGA, the Fontan procedure for PA).

Fallot's tetralogy

This is the commonest cyanotic heart defect. The four features of the tetralogy are due to deviation of the infundibular septum (area of interventricular septum between the aortic and pulmonary roots) in an anterocephalad direction:

- Pulmonary stenosis (infundibular septum obstructs the right ventricular outflow tract). Pulmonary atresia represents the severest form.
- VSD.
- Aorta overrides the VSD (receives blood from both right and left ventricles)
- RVH (due to outflow tract obstruction).

Fallot's tetralogy presents with a murmur, increasing cyanosis or hypercyanotic spells. There is a pulmonary systolic murmur and the second heart sound is single. Hypercyanotic spells occur due to infundibular spasm, which results in almost total obstruction of the right ventricular outflow tract. They may be precipitated by exertion and can be life-threatening. An affected baby becomes pale, extremely cyanosed, screams with pain and may lose consciousness. Older children characteristically squat after exertion to relieve symptoms. Squatting increases systemic vascular resistance (thus increasing pulmonary blood flow) and decreases return of desaturated venous blood from the legs.

Investigation

- Chest radiography. Oligaemic lung fields, boot-shaped heart

- ECG. Right axis deviation, right atrial hypertrophy (RAH), RVH.

Management

Emergency treatment of hypercyanotic spells includes drawing the knees up to the chest, oxygen, i.v. pain relief (e.g. morphine) and i.v. propranolol, which relieves the infundibular spasm. Oral propanolol is used as prophylaxis.

A systemic-to-pulmonary shunt is inserted to improve pulmonary blood flow in infants with marked cyanosis or uncontrolled spells. Total repair of the defect is usually carried out before school age (some centres now do a total repair as a primary procedure in the neonatal period).

Persistent fetal circulation (PFC)

PFC occurs when the pulmonary arterial pressure fails to fall following birth. This leads to persistent right-to-left shunting through the ductus arteriosus and foramen ovale. Predisposing factors include severe lung disease, birth asphyxia, polycythaemia and hypoglycaemia. Cyanosis may be progressive over the first few hours of life, with increasing respiratory distress, acidosis and hypotension. The second heart sound is loud and single. There may be differential cyanosis with a lower Po_2 in the lower body (post-ductal) than the upper body (preductal). Management includes hyperventilation and pulmonary vasodilators, e.g. tolazoline, nitric oxide.

Further reading

Anderson RH, Macartney FJ, Shinebourne EA, Tynan M. (eds) *Paediatric Cardiology.* Edinburgh: Churchill Livingstone, 1987,

Related topics of interest

Congenital heart disease – acyanotic (p. 91)
Heart failure (p. 168)
Neonatal respiratory distress (p. 235)

CONGENITAL INFECTION

Many organisms can cross the placenta to cause fetal infection, and although severe neonatal symptoms may occur, intrauterine infections are frequently subclinical at birth. Vertically acquired human immunodeficiency type 1 (HIV-1) infection is considered in a separate topic. The term 'TORCH' excludes several important pathogens, and therefore individual infections should always be considered according to the clinical presentation.

Problems

- Spontaneous abortion and stillbirth.
- Congenital malformations.
- Hydrops fetalis.
- Neonatal septicaemia.
- Non-specific illness, e.g. rash, hepatosplenomegaly.
- Long-term sequelae.

Listeriosis

Listeria monocytogenes is a Gram-positive bacterial rod which survives at low temperatures. Outbreaks have been associated with the consumption of paté, soft cheeses and coleslaw. Vertical transmission may occur transplacentally, causing septicaemia in the first 48 hours, or intranatally, usually causing meningitis and/or septicaemia during the second week of life. There may be maternal pyrexia, prolonged rupture of the membranes or meconium staining of liquor. Focal cutaneous granulomas may be seen on the baby's pharynx or back. Diagnosis is by culture of the organism from tissue, blood, CSF, surface swabs or amniotic fluid. Antibiotic treatment is with intravenous ampicillin and an aminoglycoside.

Parvovirus B19

Human parvovirus B19 is a common virus world-wide, responsible for the childhood exanthem erythema infectiosum (fifth disease, slapped cheek syndrome), arthralgia and aplastic crises in patients with red cell disorders. Twenty-five per cent of adult infections are asymptomatic. Spread is by respiratory droplet from a patient incubating the virus (before the rash appears), and viral replication then occurs in fetal haemopoietic tissue, occasionally infecting the myocardium. Spontaneous abortion, hydrops fetalis or a neonatal illness with purpura, cardiomegaly and hepatosplenomegaly may result. Diagnosis may be confirmed by antibody measurement (specific IgG and IgM) in maternal, fetal or neonatal serum or by virus identification (B19 DNA or B19 antigen) in blood, tissue or amniotic fluid.

Rubella

Rubella causes a mild viral illness in young children, but maternal rubella infection in the first 16 weeks of pregnancy may lead to fetal damage – up to 90% of pregnancies infected during the first 8–10 weeks are affected. Infants with the congenital rubella syndrome (CRS) are generally small for gestational age, with multiple congenital defects, including cataracts, retinopathy, deafness, microcephaly, congenital heart defects and bony lesions. There may be isolated sensorineural deafness. Diagnosis of congenital rubella may be confirmed by virus isolation from urine or nasopharyngeal secretions, or by the persistence of rubella-specific IgM in the infant during the first 6 months of life. There is no specific treatment. Since the introduction of MMR vaccine, cases of CRS have fallen in number. Girls between 10 and 14 years should receive rubella vaccine if not previously vaccinated with MMR.

Cytomegalovirus

Cytomegalovirus (CMV) is a herpesvirus that is usually asymptomatic in healthy individuals and, once acquired, may become latent for years. Congenital CMV infection may result from primary or recurrent maternal infection at any stage of pregnancy, and affects 3 per 1000 pregnancies in the UK, although only 10% of these will result in long-term sequelae. Clinical features in the neonate may include intrauterine growth retardation, microcephaly, sensorineural hearing loss, hepatosplenomegaly, jaundice, chorioretinitis and thrombocytopenia. Diagnosis is by virus isolation from urine, or the detection of specific IgM in neonatal serum.

Toxoplasmosis

The protozoon *Toxoplasma gondii* may be transmitted to pregnant women at various stages of its life cycle, including the ingestion of cysts in undercooked meat or oocysts via the faeces of infected kittens. Twenty per cent of pregnant women will already be seropositive following previous, often asymptomatic, infection, but a primary infection during pregnancy may infect the fetus, causing microcephaly, jaundice, hepatosplenomegaly and thrombocytopenia – the classical triad of hydrocephalus, chorioretinitis and intracranial calcification is rare. Diagnosis depends on serological techniques on maternal and neonatal serum, and parasite isolation from blood and tissue. Infection may be treated with spiramycin, alternating with pyrimethamine, sulphadiazine and folinic acid, although efficacy is unproven.

Herpes simplex

Neonatal herpes simplex virus (HSV) infection affects 2 per 1000 live births in the UK. The majority of infections are acquired perinatally from the maternal genital tract, although transplacental or post-natal spread may also be responsible. The risk of transmission is greatest if the mother has active, primary genital herpes at delivery, and Caesarean section is recommended in these circumstances. Intrauterine infection may present with scars, microcephaly and chorioretinitis at birth. Neonatal HSV infection often causes a severe disseminated infection, which may have a delayed onset of up to 3 weeks and has a mortality of 50%, even with intravenous acyclovir. Diagnosis is by electron microscopy of fluid from skin vesicles, or virus isolation from nasopharynx or conjunctivae.

Varicella zoster

Primary maternal varicella zoster virus infection in the first trimester may result in neonatal abnormalities such as cicatricial skin lesions in a dermatomal distribution, microcephaly, cerebellar hypoplasia and skeletal and eye anomalies. Primary infection in the second or third trimester or reactivated latent zoster virus carries less risk. Perinatal chickenpox, however, can cause a devastating neonatal illness with a rash and pneumonitis in babies whose mothers develop a chickenpox rash up to and including 5 days before delivery, or up to 2 days after delivery. These babies should be protected by intramuscular zoster immune globulin as soon as possible after delivery. Early neonatal treatment with acyclovir may not prevent fatalities.

Further reading

Hall SM. The diagnosis of congenital infection: three 'old' pathogens with new significance in the 1990s. *Current Paediatrics,* 1993; **3(3):** 171–9.

McIntosh D, Isaacs D. Varicella zoster virus infection in pregnancy. *Archives of Disease in Childhood,* 1993; **68:** 1–2.

Peckham CS, Logan S. Screening for toxoplasmosis during pregnancy. *Archives of Disease in Childhood,* 1993; **68:** 3–5.

Related topics of interest

Childhood exanthemata (p. 62)

Jaundice (p. 195)

CONSTIPATION

Chronic constipation in childhood may be a difficult problem, arising from a combination of physical, psychological and environmental factors. Occasionally it is a feature of serious systemic disease. Constipation implies a delay or difficulty in the passage of stool. Soiling is the inappropriate escape of stool into the clothing, usually associated with chronic constipation. Encopresis is the passage of normal stool in abnormal places, e.g. behind furniture.

Problems

- Underlying anal or myenteric plexus disease.
- Painful defecation.
- Abdominal pain and distension.
- Overflow faecal incontinence.
- Psychosocial.

Pathophysiology

Colonic propulsion delivers stool into the lower rectum, which is normally empty. However, the development of a chronically distended rectum (e.g. by avoiding defecation because of an anal fissure, or from anal stenosis) disturbs these normal sensations and their control.

Aetiology

1. Anal anomalies. Congenital problems such as anal stenosis, anterior anal ectropia, covered anus and anorectal strictures may be the reason for insufficient exit of stool.

2. Systemic disease. Constipation may be secondary to underlying disorders, e.g. hypothyroidism, hypercalcaemia, renal tubular acidosis or lead poisoning. It is a common problem in spina bifida and cerebral palsy.

3. Painful anal lesions. As the final stage of defecation is under voluntary control, the child will avoid passing stool altogether if there is associated pain or fear, e.g. anal fissure (previously wide bulky stools), rectal prolapse, perianal skin sepsis or penetrative sexual abuse. This 'retaining' behaviour leads to larger, harder stools with long-standing anal inhibition and the development of a 'megarectum'. Fear of strange or dirty lavatories may have the same effect.

4. Hirschsprung's disease. Incomplete development of ganglion cells in the plexuses of Meissner (submucosal) and Auerbach (between outer longitudinal and inner circular smooth muscle layers) results in reduced inhibition of the internal anal sphincter to rectal distension. An obstructive picture develops, and Hirschsprung's disease most commonly presents in the first week of life with neonatal intestinal obstruction. The disease is more common in boys, children with Down's syndrome or those with a family history of the condition. The prevalence is 1 in 5000. Fatal necrotizing enterocolitis is a major hazard. If there are much shorter aganglionic segments, the presentation may be that of non-Hirschsprung's megarectum (ultra-short segment).

Clinical features

Initial assessment must include the history of onset of constipation, an account of general health and diet, and in particular an account of the family's reaction to the problem. A search for signs of generalized disease and abdominal palpation to assess the degree of faecal loading is important, and, providing both parent and child are willing, a rectal examination should be performed. Often the perianal signs alone are significant (soiling, fissure, sepsis), and the rectum itself may be loaded with hard stool. In Hirschsprung's disease, a gush of faeces and air occurs on withdrawing the finger.

Investigation

- *Abdominal radiograph.* This may be used to demonstrate to doctor, parents and child the extent of faecal loading.
- *Barium enema.* Best performed on the younger, calmer infant, and on an unprepared bowel, this may demonstrate short but not ultra-short segment Hirschsprung's disease ('cone sign' of dilated bowel tapering to the narrowed aganglionic segment) and has a very limited role to define the extent of disease. Rectal biopsy is needed for definitive diagnosis.
- *Rectal biopsy.* Suction biopsy provides a small mucosal sample for histopathological diagnosis of Hirschsprung's disease – excessive acetylcholinesterase activity is seen. Rarely, results are equivocal and a full-thickness biopsy is needed. Biopsies may be combined with manual evacuation or anal stretch under general anaesthetic.
- *Anorectal manometry.* Although this method has been used to establish the presence of ultra-short segment Hirschsprung's disease, suction rectal biopsy with tissue diagnosis is much more widely available and practised. However, it may have a valuable role in the investigation of faecal incontinence, and can be used as biofeedback, helping affected children to recognize and respond to rectal sensations.

Management

If organic causes of constipation, including Hirschsprung's disease, have been excluded, successful management involves dealing with all those factors (individual to each child) which have combined to produce the problem. Bowel clearance of a hard faecal mass may be achieved by senna, best given nightly in a single dose. If unsuccessful, sodium picosulphate (Picolax) given orally, or phosphate enemas, gentle rectal washouts or microenemas may clear the mass. Manual evacuation under general anaesthetic may be kinder than repeated attempts at enemas, although sedation may be helpful for these. Once cleared, the bowel needs the regular stimulation of a soft stool and this needs to be combined with regular relaxed toilet regimes. Lactulose or docusate sodium, with high fluid and fibre intake, will help to acheive large soft stools, although the addition of a stimulant laxative (e.g. senna) is often necessary. Frequent relapses are common, and should be treated vigorously. Premature withdrawal of laxatives may be responsible, and they may be needed for a year or more.

Further reading

Clayden GS. Management of chronic constipation. *Archives of Disease in Childhood,* 1992; **67:** 340–44.

Related topic of interest

Neonatal surgery (p. 239)

CYSTIC FIBROSIS

Cystic fibrosis (CF) is an autosomal recessive disease, affecting 1 infant in 2500 in the UK. In 1989 the CF gene was identified on the long arm of chromosome 7, and the gene product is a chloride channel – the cystic fibrosis transmembrane conductance regulator (CFTR). The commonest mutation in CF is a deletion of a phenylalanine residue at position 508 (delta F508) and accounts for 70% of CF mutations in northern Europe and North America.

Problems

- Respiratory.
- Gastrointestinal.
- Hepatobiliary.
- Nutritional.
- Metabolic.
- Arthropathy.
- Reproductive.
- Psychosocial.

Clinical features

Abnormalities affect the exocrine glandular system, mainly the respiratory and gastrointestinal tracts. Neonatal presentation may be with meconium ileus (10–15% of CF cases), prolonged jaundice or as the result of prenatal diagnosis. Later in infancy, presentation may be with failure to thrive, steatorrhoea, rectal prolapse, nasal polyps or recurrent respiratory tract infections. More rarely, infants present with hyponatraemia and hypokalaemia from excessive electrolyte loss (pseudo-Bartter's syndrome), or with anaemia and hypoproteinaemia secondary to pancreatic insufficiency. In later childhood, chronic sinusitis, biliary cirrhosis, delayed puberty and diabetes mellitus may be the presenting symptoms.

Diagnosis

1. Sweat test. Pilocarpine iontophoresis is used to stimulate sweat secretion. Levels of sodium or chloride above 70 mmol/l on a sample of above 100 mg in weight are diagnostic. Values above 50 mmol/l are suspicious, and the test should be repeated. In normal samples the sodium is usually higher than the chloride, but in CF this ratio is often reversed. Tests should be performed by skilled personnel only.

2. Antenatal diagnosis. Chorionic villus sampling at 8–10 weeks' gestation or amniocentesis at 16–18 weeks allows direct gene analysis. Population screening for CF carriers has been carried out in pilot studies, and subsequent antenatal diagnosis can then be offered to couples both known to be carriers.

3. Neonatal screening. Elevated levels of immunoreactive trypsin (IRT) in the neonatal period are the result of fetal pancreatic damage *in utero*. Raised IRT levels in affected babies may be found in dried blood spots taken at the time of the Guthrie test. Blood spots may also be tested for the delta-F508 mutation.

Complications

1. Respiratory. Repeated lung infection with *Staphylococcus aureus*, *Pseudomonas aeruginosa* and *Haemophilus influenzae* leads to progressive lung disease with bronchiectasis, clubbing, abscess formation, haemoptysis and eventually cor pulmonale. Chest radiographs may show bronchial wall thickening, mottled shadows, ring shadows and consolidation. Changes over time may be assessed by the use of the Chrispin–Norman score (values are given for changes seen in the four zones of the posterolateral film). *Pseudomonas cepacia* may account for a rapid deterioration, while *Aspergillus fumigatus* may cause wheezing from allergic bronchopulmonary aspergillosis.

2. Gastrointestinal and hepatobiliary. Malabsorption with marked steatorrhoea is the main complication, resulting from pancreatic enzyme deficiency. There is malabsorption of fat-soluble vitamins. Rectal prolapse, meconium ileus, biliary cirrhosis, gall stones and portal hypertension may also complicate the disease. Meconium ileus equivalent is a partial obstruction of the terminal ileum and caecum by a faecal mass.

3. Nutrition. Poor nutritional status results from malabsorption, a high total energy requirement (infection and respiratory effort) and the anorexia and vomiting of advancing lung disease. Intakes of 120–150% of recommended daily allowances are required to maintain adequate nutrition.

4. Arthropathy. Reduced joint mobility with pain occurs in 1–2% of children with CF, as a result of specific CF arthropathy, hypertrophic pulmonary osteoarthropathy or incidental juvenile rheumatoid arthritis. Ciprofloxacin therapy may also be a factor.

5. Metabolic. Frank diabetes mellitus occurs in 2–3% of children with CF, requiring insulin. Salt depletion may occur at presentation, or as a result of hot climates.

6. Reproductive. Ninety-seven per cent of males are infertile and women are subfertile – a primary factor is the production of viscid secretions.

7. Psychosocial. The social and psychological implications of the diagnosis are enormous, with far-reaching effects on family life and expectations of the child.

Management

1. Respiratory. Physiotherapy with postural drainage and forced expiration techniques is a vital part of treatment. Inhaled bronchodilators and mucolytics may be of value. Oral antistaphylococcal drugs may be given continuously or intermittently. *Pseudomonas aeruginosa* may be treated with oral or intravenous antipseudomonal agents, depending on the clinical condition. Nebulized antibiotics, e.g. tobramycin, may achieve long-term control of *Pseudomonas aeruginosa*. Oral steroids may be needed for severe asthmatic symptoms, and for allergic bronchopulmonary aspergillosis. Assessment for heart–lung transplantation may be appropriate for those with end-stage lung disease.

2. Nutrition. A high-calorie diet with a normal or high-fat diet should be given, accompanied by enteric-coated pancreatic enzymes, e.g. Creon or Pancrease. Vitamin supplements of vitamins A, B, C, D and E should also be given, with additional vitamin K if there is liver disease. If nutrition is still inadequate, high-calorie drinks or nocturnal nasogastric or gastrostomy feeds may be added.

Other specific therapies may be needed, depending on individual circumstances, e.g sclerosant injection of oesophageal varices.

Prognosis

Effective conventional therapies now mean that the median survival for a child with CF is 40 years. New therapies may include gene therapy (restoration of the normal CF gene to affected tissues, e.g by means of viral vectors) and nebulized amiloride (a sodium channel blocker). Deoxyribonuclease reduces the viscosity of sputum, and early improvements in lung function have been shown. Nebulized recombinant α_1-antitrypsin may reduce lung inflammation. Results of large clinical trials of these new treatments may take years to emerge.

Further reading

Warner JO. Cystic fibrosis. *British Medical Bulletin,* 1992; **48:** 717–98.

Related topics of interest

Asthma (p. 29)
Chronic diarrhoea (p. 72)
Lower respiratory tract infection (p. 211)
Nutrition (p. 249)

DEMOGRAPHY AND DEFINITIONS

Demography is the study of human populations. Children under 15 years make up approximately one-third of the world's population. One-quarter (~ 350 million) live in developed countries and three-quarters (~ 1050 million) in developing countries. The UK currently has a population of 11 million children, and there are approximately 500 000 births per year. Each family doctor has approximately 2500 patients in his/her practice and, of these, about 450 are under 15 years. With the present birth rate in England and Wales of ~13.5/1000 total population, there will be between 30 and 35 births per practice each year. By far the most common paediatric disorders are those of the respiratory system. Accidents, acute respiratory illness, fevers, convulsions and acute gastrointestinal upsets are the commonest reasons for acute referral to hospital. The Office of Population Censuses and Surveys (OPCS) collects data for England and Wales on births, childhood mortality, communicable diseases, congenital malformations, hospital admissions and other surveys, which it publishes regularly. Some diseases are legally notifiable to the Communicable Disease Surveillance Centre, in particular those that are included in the immunization programme (e.g. mumps, measles, rubella, *Haemophilus influenzae* infection), acute meningitis, food poisoning and dysentery and tuberculosis. Paediatricians are also encouraged to notify certain other uncommon diseases, such as haemolytic uraemic syndrome, so that epidemiological and outcome data can be collected.

Mortality rates

The OPCS has defined mortality rates in different age groups. The neonatal, post-neonatal and infant mortality rates are three important indicators of child health. Neonatal and perinatal mortality rates provide an indirect measure of the quality of obstetric and neonatal services. Approximate current rates for England and Wales are given in brackets.

Stillbirth rate (5/1000) The number of infants born after 26 weeks who show no signs of life per 1000 *total* births.

Perinatal mortality rate (8–9/1000) The number of stillbirths and deaths in the first week per 1000 *total* births. Principal causes are low birth weight and congenital malformations. The rate is four and a half times higher for multiple births than for singletons.

Neonatal mortality rate (4.5–5/1000) The number of deaths in the first 27 days of life per 1000 *live* births. The neonatal mortality rate can be further subdivided into early (deaths in the first 7 days of life) and late (deaths occurring from the 8th to the 27th day inclusive). Eighty per cent of neonatal deaths occur in the first week, the main causes being congenital abnormality, respiratory distress and prematurity. Neonatal deaths account for 40% of all childhood deaths, the biggest risk factor being low birth weight. Increased rates are also associated with maternal age under 20 years or over 35 years, parity of three or more and lower social class. The place of maternal birth is also a factor, for example mothers born in Pakistan have a higher rate than mothers of the same ethnic group born in this country. Rates vary between different regions of the country with the lowest figure in 1987 in East Anglia, and the highest in Yorkshire, but the relationship with poverty is not exact. Infants classified as illegitimate have a higher rate at about 6/1000.

Post-neonatal mortality rate (3–4/1000)

The number of deaths between 28 days and 1 year of life per 1000 *live* births. The major causes are sudden infant death syndrome (SIDS), congenital abnormalities and respiratory infections. A small number are late deaths due to low birth weight/prematurity. The rate of SIDS was 2/1000 in 1987, but has recently halved following changes in sleeping position advice. As with neonatal mortality, deaths in the post-neonatal period are related to age, parity, social class and place of birth of the mother, and birth weight and legal status of the infant. Deaths in this age group show a more pronounced seasonal pattern than at any other time in childhood, with more deaths in the winter, mainly attributable to SIDS and respiratory infections.

Infant mortality rate (IMR, 7–8/1000)

The number of deaths in the first year of life per 1000 *live* births. Some Third-World countries have rates in excess of 150/1000. Again, one of the major causes in the UK is congenital abnormality, including congenital heart disease. The IMR has fallen steeply from over 15/1000 in 1975 mainly because of a fall in neonatal deaths.

Mortality in later childhood (~25/100 000 children 1–14 years)

Accidents account for one-quarter of deaths between the ages of 1 and 14 years, with road accidents accounting for the majority. On average, three children die in accidents every day and ~10 000 are permanently disabled each year. Malignancy is the second commonest cause of death in this age group. The remainder include children with inherited disorders (e.g. cystic fibrosis) and acute infections.

Further reading

Forfar JO. Demography, vital statistics, and the pattern of disease in childhood. In: Campbell AGM, McIntosh N, eds. *Forfar and Arneil's Textbook of Paediatrics,* 4th edn. Edinburgh: Churchill Livingstone, 1992; 1–17.

Madeley RJ. Recent trends in infant, neonatal and postneonatal mortality rates. *Current Paediatrics*, 1991; **1:** 49–52.

Office of Population Censuses and Surveys Reference Series (various years). Mortality statistics (DH1, DH2, DH3, DH6), birth statistics (FM1), congenital malformation statistics (MB3). London: HMSO.

Related topics of interest

Prematurity (p. 271)
Sudden infant death syndrome (p. 313)

DEVELOPMENTAL ASSESSMENT

Developmental examination involves a system of procedures (history, observations, examination and tests) which are designed to establish the stage of the child's development and identify any deviations from the normal pattern. Developmental assessment involves detailed, expert multidisciplinary investigation of a suspected developmental disorder. A knowledge of the range of normality for each developmental milestone is essential for developmental screening. Briefly, essential milestones are:

6 weeks

- Holds head transiently up to the plane of the trunk in ventral suspension.
- Lies prone with a flat pelvis.
- Smiles back in response to mother.
- Follows object from side to midline.

3 months

- Head held for prolonged period above the plane of the body.
- If pulled to sitting position, minimal head lag only.
- Holds object if it is placed in the hand.
- Turns to sound.
- Hand regard (3-5 months).

6–7 months

- Sits without support by 7 months.
- Rolls prone to supine (6 months), supine to prone (7 months).
- Supports weight when held in standing position.
- Transfers objects from hand to hand.
- Feeds self with biscuit and holds own bottle.
- Babbles of two syllables.

9 months

- Sits steadily unsupported.
- Pulls up to stand, holding on to furniture.
- Compares two cubes, bringing them together.
- Pincer grasp, e.g. picks up currant between index finger and thumb.

1 year

- Walks with one hand held, or creeps 'like a bear' on hands and feet.
- Releases toys into mother's hand.
- Says 2–3 words with meaning, but understands many more.

18 months

- Walks upstairs if hand is held.
- Domestic mimicry.
- Builds tower of 3–4 bricks.
- Obeys two simple commands.
- May be dry by day.

2 years

- Walks up and down stairs alone.
- Draws circular and vertical strokes with a pencil.
- Builds tower of 6–7 bricks.
- Requests food and toilet needs.
- Forms two spontaneous three-word sentences.
- Watches other children at play.

3 years

- Rides a tricycle.
- Undoes buttons, mainly dressing self.
- Copies circles on paper, imitating crosses; may draw a man.
- Builds tower of nine bricks.
- Asks numerous questions. Joins sociable play.

4 years

- Skips. Goes up and downstairs with one foot per step.
- Does up buttons.
- Copies cross on paper.
- Self-caring for toilet needs.
- Plays 'make-believe' games.

History

A developmental history must explore parental concerns about the child's development to date. Milestones which have been achieved should be noted. A full family history must be taken, and should include details of vision, hearing and special educational needs. Maternal age and well-being during the pregnancy may be relevant, and full details of the birth and early neonatal period should be obtained. Significant illnesses or accidents may have occurred since birth. A thorough social history will provide information about the level of stimulation and learning opportunity in the child's home environment.

Examination

Physical examination should be deferred until after the neurodevelopmental assessment, but rapid observation may detect abnormalities of gait and movement, characteristic facies, or eye abnormalities. Later, the examination should include measurements of growth, including head circumference, palpation of the anterior fontanelle and examination of the heart, chest, abdomen, testes and hips.

Neurodevelopmental examination

Formal neurological assessment is best combined with developmental examination in children of most ages. Testing for the persistence of primitive reflexes should be included, where relevant. Cranial nerve function and muscle tone should be tested.

Other tests will depend on the age of the child. Usually only limited equipment is necessary, but should include such items as a set of 1-inch cubed bricks, small pellets of paper, a ring dangling on a thread, and a small handbell. Much information can be obtained by observation of the quality of performance of the various tasks, and the associated vocalization.

The Working Party on child health surveillance (Hall Report) found no clear justifications for routine formal developmental assessment for all children, recommending the evaluation of history, parental concerns and simple milestones only. Referral for expert opinion should be made if there are doubts about normal progress. However, specific tests are in general use for developmental assessment.

- The Denver developmental screening test is widely used for screening normal children.
- Bayley scales of infant development give separate scores for mental and motor development and include an observation of behaviour.
- Griffith's test is widely used for those children in whom an abnormality is found during developmental screening. It gives separate scores for locomotor, personal/social, hearing and speech, hand–eye coordination and performance skills, and a sixth subtest, practical reasoning, is included for 3- to 8-year-olds. Each subtest gives a developmental quotient (DQ) which is averaged into a general quotient (GQ).
- The Mary Sheridan 'from birth to 5 years' scheme is based on the detailed concepts and tests of Gesell and is widely used in the UK.

Clinical assessments are preferable to scales and scores in the serial assessment of handicapped children.

Further reading

Hall DMB, ed. *Health for all Children. A Programme for Child Health Surveillance.* Oxford: Oxford University Press, 1991.
Illingworth RS. *The Development of the Infant and Young Child,* 9th edn. Edinburgh: Churchill Livingstone, 1987.
Pollak M. *Textbook of Developmental Paediatrics.* Edinburgh: Churchill Livingstone, 1993.

Related topics of interest

Cerebral palsy (p. 56)
Developmental delay and regression (p. 115)
Hearing and speech (p. 165)
Vision (p. 325)

DEVELOPMENTAL DELAY AND REGRESSION

Developmental examination may detect a pattern of development which varies from the expected result. The progression of milestones may be normal in sequence, but delayed in time when compared with an average child of the same age, or the sequence may be distorted and abnormal. Regression, with loss of previously acquired skills, is another abnormal pattern.

Developmental delay may affect:

- Motor development.
- Speech and language.
- Global development, i.e. all parameters affected.

For a discussion of the developmental history and neurodevelopmental examination, see Developmental assessment, p. 111.

Causes of delayed motor development

- Normal or familial variation.
- Previous chronic illness.
- Cerebral palsy.
- Neuromuscular disorders, e.g. Duchenne muscular dystrophy.
- Orthopaedic problems, e.g. congenital dislocation of the hip.
- Emotional neglect.

Disorders of speech and language are discussed in a separate topic, Hearing and speech (p. 165). Hearing defects should always be excluded, and the overall pattern of development assessed, to exclude global developmental delay.

Causes of global developmental delay

Milestones may be delayed, but may later catch up, e.g. in a child with a previous chronic disease or an emotionally deprived child who subsequently has a secure and loving foster home. However, many globally delayed children will have learning difficulties of varying degrees. In many cases a specific aetiology will not be found, but it is important to look hard for a specific cause, for reasons of prognosis and genetic counselling. A cause is more often found in those with severe learning difficulties (IQ below 55), whereas those with mild to moderate learning difficulties (IQ 55–75) frequently represent the lower end of the normal spectrum.

1. Chromosomal abnormalities. Down's syndrome is the commonest chromosomal abnormality to cause developmental delay, followed by the fragile X syndrome.

Many rarer abnormalities associated with specific syndromes are now being recognized.

2. *Structural brain abnormalities.* Microcephaly, hydranencephaly, agenesis of the corpus callosum and abnormalities of the cortical gyri may be responsible for global developmental delay.

3. *Specific syndromes.* Numerous specific syndromes associated with global delay have been recognized, e.g. Cornelia de Lange syndrome, Smith–Lemli–Opitz syndrome. The neurocutaneous syndromes (tuberous sclerosis, Sturge–Weber, neurofibromatosis) may also be responsible.

4. *Congenital infection.* Congenital rubella, toxoplasmosis, cytomegalovirus and herpes simplex infection may cause developmental delay in association with other defects, e.g. cataracts, intracranial calcification.

5. *Inborn errors of metabolism.* Phenylketonuria and congenital hypothyroidism should now be excluded by the routine neonatal screening programme. Other individual inborn errors are rarer, e.g. the mucopolysaccharidoses, maple syrup urine disease, etc.

6. *Cerebral palsy.* Some children with cerebral palsy will have global delay.

7. *Post-natal causes.* Post-natal brain insults such as severe head injury, intracranial haemorrhage, meningitis or encephalitis may cause global developmental delay in a child with previously normal milestones.

Developmental regression

The loss of previously acquired skills is an ominous developmental finding, and may be indicative of a progressive neurometabolic brain disease such as Batten's disease, metachromatic leucodystrophy.

Investigation of global delay

Selected investigations may be necessary:

- Chromosomal analysis.
- Imaging of CNS – by CT or MRI scan – to exclude structural abnormalities.
- Urine tested for virus excretion (rubella, cytomegalovirus), mucopolysaccharides, amino acids.

- Thyroid function tests.
- Serological tests for congenital infections.
- Creatine phosphokinase in boys.

Management

Very few causes of global developmental delay are amenable to specific treatment. Parents and carers need much help and repeated discussions in coming to terms with this diagnosis. The identification of a specific pathology may provide a prognostic guide, and allow genetic counselling. Practical help may be needed with feeding, sleeping and behavioural difficulties, and the provision of respite care. Financial support may be needed, e.g. the Disability Living Allowance, the Joseph Rowntree Memorial Trust. Special educational provision must be considered at an early stage, based on a Statement of Educational Need.

Further reading

Pollak M. *Textbook of Developmental Paediatrics.* Edinburgh: Churchill Livingstone, 1993.

Related topics of interest

Cerebral palsy (p. 56)
Developmental assessment (p. 111)
Hearing and speech (p. 165)

DIABETES MELLITUS

Childhood diabetes mellitus is almost invariably insulin dependent (type I, IDDM). Insulin deficiency results from irreversible damage to the β-cells of the pancreatic islets of Langerhans. The incidence increases with distance from the equator, is higher in whites than in non-whites and peaks at adolescence (1 in 1200 school-children). The aetiology remains unclear but is likely to involve both genetic predisposition and environmental factors.

Problems

- Hypoglycaemia.
- Ketoacidosis.
- Cerebral oedema.
- Family and social disruption.
- Long-term complications (retinopathy, nephropathy, accelerated atherosclerosis, peripheral neuropathy).

Aetiology

1. Primary

(a) Genetic factors

- Increased incidence in siblings (15–50 times normal) and parents. However, only 10% of affected children have a diabetic first-degree relative.
- Five times more common in monozygotic twins than in dizygotic twins.
- Association with HLA-DR3 and -DR4.

(b) Environmental factors

- Seasonal variation in incidence (autumn/winter peaks).
- Case clustering (geographical variation).
- Association with viral infections, e.g. mumps, coxsackie B4. Destruction of the β-cells may take place over several years with acute decompensation precipitated by a virus or other toxin.

(c) Autoimmune factors

- Associated with autoimmune diseases of thyroid, adrenal and parathyroid.
- Cytotoxic islet cell antibodies often found at diagnosis.

2. Secondary

- Cystic fibrosis.
- Genetic syndromes, e.g. Prader–Willi.
- Cushing's syndrome.

Clinical features

The majority of children are diagnosed before the onset of severe ketoacidosis. Characteristic features include thirst, polyuria, weight loss, poor appetite and malaise, developing over days or weeks. Enuresis may recur in a previously dry child. Occasionally a child presents more acutely with rapid onset of ketoacidosis progressing from vomiting, dehydration and hyperventilation to shock, drowsiness and coma over a few hours. Abdominal pain may be a prominent feature, resulting in admission to a surgical ward with an 'acute abdomen'.

Examination should assess:

- Degree of dehydration.
- Level of consciousness.
- Pulse, blood pressure.
- Respiratory rate – Kussmaul respiration (deep sighing). Breath may smell sweet (acetones).
- Sites of possible infection which may have precipitated acute decompensation.

Investigation

- Urinalysis. Glycosuria strongly suggests the diagnosis. Ketones may be present. Other causes of glycosuria include renal tubular disorders (normal blood glucose).
- Random blood glucose > 11 mmol/l confirms the diagnosis. Glucose tolerance tests are rarely indicated in children.
- U&E. Hyperglycaemia results in polyuria and dehydration. Hyponatraemia results from osmotic diuresis and shift of water from within cells to the extracellular fluid. Despite a total body deficit, the serum potassium is usually normal at presentation because of acidosis.
- Acid–base status. Metabolic acidosis is due to excess lipolysis and production of ketones.
- FBC. White count may be raised even in the absence of infection.
- Blood cultures, urine culture, throat swab ± chest radiography.

Initial management (not ketoacidotic)

In most centres children are admitted to hospital for a few days to allow education of the family and child about diabetes, and to commence insulin injections. Simple practical information should be given about insulin, diet, injection techniques and the symptoms of hypoglycaemia. Diabetic liaison nurses and dietitians are invaluable members of the team. This education continues at home and in out-patients. Initial insulin requirements are ~ 0.5 units/kg/day subcutaneously given as a twice-daily regimen of short- and medium-acting insulins, two-thirds being given before breakfast and one-third before the evening meal. Requirements may be increased by ketoacidosis at presentation, or intercurrent illness, and may be reduced if the diagnosis is made early, with only mild metabolic decompensation.

Ketoacidosis

Diabetic ketoacidosis (DKA) is an emergency which requires prompt recognition and treatment. It may be the presenting feature of IDDM or occur in a known diabetic during an intercurrent illness, or when insulin is not given (in error or deliberately). A common mistake is to stop insulin in children who are not eating because they are unwell or vomiting. In this situation they often need more insulin.

- Treat shock with 10–20 ml/kg of plasma.
- Fluids. Intravenous 0.9% saline initially (change to 0.45% saline/2.5% dextrose when blood sugar <11 mmol/l). Calculate fluid deficit (usually 10–15% dehydrated) and normal requirements and replace at a constant rate over 24 hours. Insulin promotes the uptake of potassium into cells so monitor carefully when treatment is commenced and add potassium to fluids. The metabolic acidosis will resolve with correction of dehydration and hyperglycaemia so bicarbonate is rarely needed.
- Insulin. Intravenous infusion commencing at 0.1 units/kg/h.
- Nil by mouth for first 12 hours at least. A nasogastric tube should be passed if vomiting is profuse or gastric dilatation is present.
- Urinary catheterization is unnecessary unless the patient is comatose, a distended bladder is palpable or no urine has been passed within 4–6 hours of commencing treatment.
- Antibiotics if a focus for infection is identified.

Cerebral oedema

There are 6–11 deaths per year from DKA in children under the age of 19 years in the UK. Cerebral oedema is the most important cause of mortality and morbidity. It develops a few hours after commencing treatment in the context of apparent biochemical improvement. It is difficult to predict those at greatest risk, and it may occur in seemingly mild DKA. Rapid fluid replacement and free water overload may contribute, but the antecedents to the development of cerebral oedema are still unclear.

Hypoglycaemia

See p. 180.

Long-term follow-up

1. Maintenance of good control. Monitoring is best done by finger-prick blood tests. Urine tests may be used but these are crude, with individual variation in the level at which glucose spills into the urine. Glycosylated haemoglobin (HbA1) gives an index of control over a longer period (normal up to 8%). The debate on whether improved control reduces the risk of long-term complications is ongoing, but accumulating evidence suggests that optimal control should be promoted. However, this must be balanced against avoidance of recurrent hypoglycaemia and maintenance of an acceptable lifestyle.

2. Growth and development. Physiological and behavioural adjustments in adolescence and puberty lead to changes in insulin requirements. Relative insulin resistance is due to increased growth hormone. Multiple injection regimens with pen injection devices are often useful.

3. Long-term complications. Screening should include annual retinal examination and urine for proteinuria. Microalbuminuria may be an early predictor of nephropathy. Smoking should be discouraged.

Further reading

Dunger DB. Diabetes in puberty. *Archives of Disease in Childhood,* 1992; **67:** 569–70.
Johnston DI. Management of diabetes mellitus. *Archives of Disease in Childhood,* 1989; **64:** 622–8.

Related topics of interest

Coma (p. 85)
Hypoglycaemia (p. 180)

DROWNING AND NEAR-DROWNING

Drowning is the third most common cause of childhood accidental death in the UK (following road traffic accidents and burns). Over the last decade, drowning accidents accounted for an average of 72 childhood deaths per year (0.7 per 100 000 children). Many deaths occur at the scene of the immersion, and so the near-drowned child is an unusual presentation in the accident and emergency department. However, many of these children will survive normally, and some will survive with significant neurological deficit. It is therefore essential that there is a structured approach to the primary prevention of immersion incidents and to the initial resuscitation of near-drowned children.

Problems

- Cardiorespiratory arrest.
- Respiratory distress.
- Cardiac arrhythmias.
- Hypothermia.
- Electrolyte disturbance.
- Infection.
- Associated injuries.
- Raised intracranial pressure.

Specific problems will be determined by the duration of immersion, the type of water (fresh or salt) and resuscitation at the scene.

Potential settings for immersion accidents

1. Bathroom. Children under 5 years old may be left unattended, or with an older child. Child abuse may present in this way.

2. Gardens. Often the garden is unfamiliar to toddler and parent, e.g. homes of relatives or commercial garden centres, where ponds, pools and ditches become the sites of the immersion accident.

3. Swimming pools. Drowning is more common in private pools, often when an unsupervised toddler wanders into an unfenced pool. Non-weight-bearing covers may also play a part in these incidents.

4. Public pools. The Health and Safety at Work Act 1974 insists on a high level of supervision in public pools, and mortality is low in this setting. Deaths tend to be in older children.

5. Open waterways and coast. Deaths tend to be in older boys, often swimming unsupervised, or playing on rafts or in boats. Poor swimming proficiency in the challenge of cold open waters, lack of adequate life-jackets and a tendency towards risk-taking behaviour may contribute.

Clinical assessment and management

If the patient has been pulled pulseless from the water, immediate cardiorespiratory resuscitation should be commenced at the scene, clearing as much water from the airways as possible. These measures should be continued in hospital with advanced life support and mechanical ventilation, aiming to rewarm the hypothermic child slowly; resuscitative measures should not be discontinued until the child's core temperature is normal and constant. Thorough examination should seek signs of associated injuries, especially of the head and neck. Broad-spectrum antibiotics may be indicated, especially if the water was heavily contaminated. Measures to control rising intracranial pressure should be instituted (intracranial pressure monitoring may be of great value), and seizures should be controlled. Regular measurement of acid–base and serum electrolytes is essential, as both fresh and salt water may cause gross electrolyte disturbance. Arrhythmias should be treated appropriately. The child who has made a rapid recovery after resuscitation should continue to be monitored closely in hospital for at least 24 hours as secondary respiratory complications may occur.

Investigations

- FBC with reticulocyte count.
- U&E, creatinine, glucose.
- Arterial blood gas and acid–base estimation.
- Chest radiograph.
- Infection screen, especially of initial endotracheal tube aspirate.
- Continuous monitoring of ECG, intracranial pressure (if available) and pulse oximetry.

Prevention

A national system of data collection of immersion incidents is important for planning preventative measures. In addition to swimming lessons, education about water safety should be part of the school curriculum. It should also be part of the child surveillance plan. Fences and self-locking gates around domestic pools and grids over garden ponds may save lives. Additional legislation may be necessary to achieve some of these measures.

Further reading

Kemp AM, Sibert JR. Outcome in children who nearly drown: a British Isles study. *BMJ,* 1991; **302:** 931–3.
Kemp AM, Sibert JR. Drowning and near drowning in children in the UK: lessons for prevention. *BMJ,* 1992; **304:** 1143–6.
Pearn J. The urgency of immersions. *Archives of Disease in Childhood,* 1992; **67:** 257–8.

Related topics of interest

Child abuse (p. 59)
Cardiopulmonary resuscitation (p. 52)

DYSMORPHOLOGY AND TERATOGENESIS

Dysmorphology is the study of malformations and birth defects. Abnormal development may be due to a chromosomal abnormality, a single gene defect or a teratogen, or may be multifactorial. Disruption or deformation of a normal fetus may occur as a result of amniotic bands or oligohydramnios, and abnormal intrauterine posture due to neuromuscular disorders may lead to deformities of the legs and feet (e.g. talipes). A syndrome is a recognized pattern of abnormalities that have a single cause (e.g. Down's syndrome). Malformations may also occur as a result of a sequence of events following a single initiating abnormality (e.g. Pierre Robin sequence due to micrognathia) or occur in associations (e.g. CHARGE association). Recognition of patterns of multiple congenital abnormalities may help in identifying the aetiology and is important in counselling parents about prognosis, management and risk of recurrence. The majority of single malformations have a multifactorial inheritance and a low recurrence risk. Multiple malformations, particularly when associated with learning difficulties, are more frequently due to a single gene defect or a chromosomal abnormality and the recurrence risk is higher.

Assessment of a dysmorphic infant

A careful history should include details of the pregnancy, exposure to possible teratogens, parental age, consanguinity and family history. All the abnormalities should be documented and clinical photographs taken where possible. Further investigations may include chromosomal analysis, DNA studies, metabolic investigations and radiology. Numerous congenital malformations and syndromes have been described, and many are extremely rare. Specialist texts and computer programs assist in the differential diagnosis.

Miscarriage or stillbirth may be due to congenital malformation, and it is important to examine all abortuses and stillborn infants carefully. If congenital abnormalities are found, cardiac blood samples and skin biopsies can be taken for biochemical and genetic studies to try to identify the defect so that parents can be counselled about the recurrence risk and the availability of prenatal tests in future pregnancies.

Teratogens

Maternal illness, infection or ingestion of certain drugs during early pregnancy can lead to congenital malformation. The most widely known teratogenic drug is the antiemetic thalidomide, which damaged about 10 000 babies world-wide in the early 1960s before it was withdrawn. Other known teratogens include:

- Drugs, e.g. alcohol, anticonvulsants (sodium valproate, phenytoin), anticoagulants (warfarin).

- Maternal disorders, e.g. diabetes, phenylketonuria.
- Intrauterine infection, e.g. toxoplasmosis, rubella, cytomegalovirus.
- Environmental chemicals, e.g. organic solvents.
- Radiation.

Fetal alcohol syndrome

Alcohol is now the commonest teratogen ingested during pregnancy. Even a moderate alcohol intake is associated with an increased risk of spontaneous abortion and mild growth retardation. Fetal alcohol syndrome is seen in babies born to chronic alcoholic mothers. Typical features include a characteristic facial appearance (short palpebral fissures, smooth philtrum, thin upper lip), prenatal and post-natal growth retardation, microcephaly and mild to moderate learning difficulties.

Associations

An association is a combination of abnormalities which occur together more often than expected by chance and the names are often acronyms of the components, e.g. CHARGE association = colobomas of the eye, heart defects, choanal atresia, mental retardation, growth retardation and ear anomalies.

Further reading

Baraitser M, Winter RM. Genetics and congenital malformations. In: Roberton NRC, ed. *Textbook of Neonatology*, 2nd edn. Edinburgh: Churchill Livingstone, 1992; 743–75.

Lyons Jones K. Dysmorphology – the approach to structural defects of prenatal onset. In: Behrman RE, ed. *Nelson Textbook of Paediatrics*, 14th edn. Philadelphia: WB Saunders, 1992; 301–4.

Related topics of interest

Antenatal diagnostics (p. 23)
Chromosomal abnormalities (p. 68)
Congenital infection (p. 98)

ECZEMA

Atopic eczema affects 3% of children under 5 years old, most commonly from 3 months to 2 years. The skin changes include intraepidermal oedema, which is followed by inflammation and intense itching of the skin. The skin is often erythematous, and may ooze and crust, with later changes of lichenification or alterations of pigmentation. Up to 80% of children have a positive family history of atopic conditions. A minor defect of cell-mediated immunity and suppressor T-cell numbers is thought to be a factor in pathogenesis. Rarely, eczema is a feature of another disorder, e.g. Wiskott–Aldrich syndrome, phenylketonuria.

Problems

- Itching.
- Cosmetic disfigurement.
- Secondary infection.

Clinical history

The child typically presents with affected areas on the scalp, face and trunk which itch and scale. The extensor surfaces of the limbs and the dorsal areas of hands and feet are often involved. It is important to know the full medical history, any known exacerbating factors such as foods, contact with pets or pollen and the family history of atopic disease. It must be established which treatments have previously been used on the skin. It is also useful to establish the nature of the parents, and child's concerns, e.g. sleepless nights from itching, and their expectations from treatment.

Examination

Height and weight should always be documented, and a general physical examination performed. Affected areas of skin may show erythema, vesicles, lichenification or crusting. Painless lymphadenopathy is very common. The skin may exhibit white dermographism, with blanching of the skin after light pressure, and an extra infraorbital fold, the Dennie–Morgan sign, is often seen. A search should be made for signs of infection (pustules, crusting and weeping).

Investigation

- IgG, RAST and immunology. IgE is raised in 80% of cases, and is of little practical help. There may be positive radioallergosorbent (RAST) tests and an eosinophilia. Detailed studies of immune function are sometimes indicated.
- Skin swabs. *Staphylococcus aureus* is invariably cultured from eczematous skin, but swabs often provide useful antibiotic sensitivities. Lancefield group A β-haemolytic streptococcus is a significant cause of infection. If lesions have a vesicular component and eczema herpeticum is suspected, viral swabs should be performed.
- Biochemistry. In severe eczema with oedema, hypoalbuminaemia may be found.

1. Itching and scratching. Nails should be kept short and clean, and mittens may be helpful at night. Splints are not indicated. Night-time sedatives are often very helpful, e.g. trimeprazine, promethazine, chlorpheniramine.

2. Dryness. Every attempt should be made to keep the skin moist with emollients, e.g. aqueous cream and paraffin-based creams. Ordinary soap is often drying and irritant and should be avoided. Bath oils may also be used, or the emollient may be applied before the bath. Coal tar or zinc bandages are of limited value for lichenified eczema.

3. Inflammation. If inflammation is severe, topical treatment with steroid ointments or creams is indicated. Low steroid potency preparations should be used, e.g. 0.5–1% hydrocortisone the lowest strength being used in facial areas. Occasionally a more potent strength is needed for short periods on selected areas of skin. Side-effects include skin thinning, striae and secondary infection. In uncontrolled cases, oral prednisolone is used. As short stature is not uncommon in severe eczema, effects of uncontrolled skin inflammation must be weighed against steroid side-effects.

4. Infection. Secondary infection may require systemic treatment or local agents, depending on the clinical picture and the organisms cultured. Oral flucloxacillin or a cephalosporin may be appropriate. Eczema herpeticum is a severe systemic illness in eczema sufferers which follows a primary infection with herpes simplex. Aggressive supportive therapy and intravenous acyclovir is indicated.

5. Dietary restriction. For a few children, dietary elimination of dairy products produces an improvement in their eczema. Neither skin tests nor RAST are of practical help. More restrictive in-patient elimination diets have also been tried with some improvement for severe disease. Expert dietary advice is essential, as protein, energy, calcium and iron deficiencies may occur.

6. Other treatment measures. Trigger factors should be avoided. There is no convincing evidence that exclusive breast-feeding is protective, although dietary factors in the mother may be relevant. Gamolenic acid in evening primrose oil has been used in recent years, and there has also been widespread interest in Chinese herbal preparations.

Further reading

David TJ. Management of atopic eczema. In: David TJ, ed. *Recent Advances in Paediatrics,* Vol. 12. Edinburgh: Churchill Livingstone, 1994: 187–204.

Devlin J, David TJ, Stanton RHJ. Elemental diet for refractory atopic eczema. *Archives of Disease in Childhood,* 1991; **66:** 93–9.

Related topics of interest

Allergy (p. 17)
Asthma (p. 29)
Blistering conditions (p. 41)
Rashes and naevi (p. 285)

ENURESIS

Involuntary emptying of the bladder during sleep is a common problem. Most children are dry at night by the age of 3 years, but about 10% still wet the bed regularly at 5 years, approximately 5% still wet at 10 years and 1% still wet at 15 years. Enuresis is more common in boys and there is often a family history of bed-wetting. Only 1–2% will have an underlying organic problem.

Problems

- Inconvenience and expense. Frequent laundry, damage to mattress.
- Psychological. Embarrasment, fear of staying away from home, parent–child conflict.

Aetiology

- Idiopathic (98–99%). Delay in maturation of bladder control.
- Urinary tract infection (UTI).
- Congenital abnormality of urinary tract.
- Polyuria.
- Neurogenic bladder.
- Mental subnormality.
- Nocturnal epilepsy.
- Constipation.

Clinical features

The majority of cases are due to delay in maturation of bladder control. There is often a family history of enuresis (70%), and it is more common in lower socioeconomic groups. Many cases are exacerbated by faulty management of toilet training, and emotional problems or psychological upset are common. Most children presenting with enuresis have never been dry (primary enuresis), but recurrence of wetting after being dry for months or even years (secondary enuresis) may occur during periods of psychological upset, e.g. parental separation, change of school. A few cases of secondary enuresis are due to an organic problem, e.g. UTI, polyuria secondary to diabetes mellitus, constipation. A history of constant dribbling of urine day and night suggests an underlying urinary tract abnormality (e.g. ectopic ureter opening into the urethra) or a neurological problem (e.g. diastematomyelia, sacral agenesis). Careful examination of the spine, genitalia, anus and lower limbs is essential. Radiological investigation of the renal tract is rarely indicated, but features such as hypertension, a family history of renal tract disease (other than enuresis), dribbling incontinence, abnormal lower limb neurology or abnormal urinalysis would prompt further investigation.

Investigation

- Urinalysis. Glucose, protein, blood, specific gravity.
- Urine. Culture.

Management

1. General measures. Ensure that the child does not drink excessively just prior to bed-time. Lifting and putting the child on the toilet last thing at night when the parents are going to bed is often successful in preventing wetting later in the night and may help with bladder training. Smacking or ridiculing the child for wetting the bed is detrimental. Excessive anxiety or conflict over toilet training is unhelpful. Some parents have unrealistic expectations as to when a child should be dry (e.g. by the age of 2 years) and need reassurance that wetting at this age is not abnormal.

2. Reward systems. Encouragement and positive reinforcement for dry nights, e.g. with star charts, is helpful in bladder training. The pad and buzzer method is sometimes helpful in older children. A pad under the sheet, when wet, completes an electrical circuit and causes a buzzer to ring. The buzzer should wake the child as soon as he starts to pass urine and he can then go to the toilet to finish voiding. Children are thought to become more aware of the sensation of a full bladder using this method, and it can be used in combination with a star chart. The major complaint from parents about these buzzers is that they wake other children in the same room and they not infrequently fail to wake the urinating child.

3. Drugs. Drug therapy is not appropriate for children under 7 years and should be reserved for when simple measures, such as bladder training and alarm systems, fail. Tricyclic drugs (e.g. imipramine) are sometimes used because of their antimuscarinic effects. These work temporarily, resulting in complete dryness or a significant improvement in about one-third of children, but relapse is common on withdrawal of treatment. Tricyclics may be associated with behaviour disturbances, and the risks of overdosage should be borne in mind. Treatment should not exceed 3 months. Desmopressin is an antidiuretic agent given by nasal spray. This is reported to help 30–50% of children, but again the relapse rate is high. It is useful as an intermittent treatment for use when the child particularly wishes to avoid wetting the bed, e.g. staying with friends, going away on a school trip.

Outcome

Whatever method is used, most children will improve with time, and the majority will stop bed-wetting as they get older. Parents and children should be reassured that this is a common condition and that the natural outcome is good. Anxiety and conflict about bed-wetting will only exacerbate the situation.

Related topic of interest

Urinary tract infection (p. 322)

FAILURE TO THRIVE

Failure to thrive is a symptom which is specific to childhood, indicating a failure to gain weight at the expected rate. Diagnosis is based on serial measurements over time, which may show weight loss, static weight or height, or serial measurements which cross the centiles to a lower position. The term is usually reserved for problems of infant growth (generally under 2 years of age). The term is not a diagnosis in itself, and multiple causes, both organic and non-organic, may interreact to produce the undernutrition which is primarily responsible.

Causes of failure to thrive

In many cases, a combination of factors is responsible, and community-based studies suggest that no more than 10% of cases have an organic basis. Broad categories of underlying causes are:

1. Defective calorific intake. Poor breast-feeding techniques may lead to the consumption of inadequate quantities of high-fat hindmilk. Appetite may be suppressed by chronic disease or drugs. Insufficient food may be offered or behavioural problems may occur at meal-times. Disordered swallowing or sucking mechanisms, e.g. cerebral palsy, may be responsible.

2. Abnormal nutrient losses. Persistent diarrhoea or vomiting may prevent adequate weight gain, as the actual intake may not be absorbed, e.g. gastro-oesophageal reflux, cystic fibrosis, coeliac disease.

3. Chronic systemic disease. Chronic diseases of cardiovascular, respiratory, renal, hepatic or central nervous system origin may cause failure to thrive. Endocrine, metabolic and infective disease must also be considered.

4. Genetic disease. Chromosomal disorders, skeletal dysplasias and dysmorphic syndromes may cause symmetrical intrauterine growth retardation (child is also short with a small head circumference) and weight then usually follows below but parallel to normal centiles.

Clinical assessment

History

Details of the heights of the child's parents and siblings should be noted, and a full history of the pregnancy, birth and neonatal period should be taken. Parent-held or clinic measurements may provide information about post-natal growth. Enquiry should be made about any systemic symptoms which may be relevant, e.g. vomiting, diarrhoea, dyspnoea, polyuria, etc. A full feeding history should establish the types and quantities of food and drink offered, and the child's responses to these. Detailed assessment of nutritional intake should then be made by a dietitian. A full social history is important, and may be supported by reports from the primary health care team or social worker.

Examination

Measurements of weight, height, sitting height and head circumference should be made personally whenever possible, and should be used in conjunction with any previous measurements to assess the velocity of weight and height gain. Correction should be made for prematurity, and a mid-parental height prediction should be made. Mid upper-arm circumference is a useful indicator of muscle bulk, and skinfold thickness over triceps or subscapular muscles is an indicator of body fat. Full physical examination may reveal dysmorphic features, or features of specific organ system disease, although in many cases examination will be unremarkable. There may be signs of physical abuse or neglect.

Investigation

Investigations may be preceded by a trial of feeding, which may establish that the child can gain weight if given an adequate dietary intake. Hospital admission is occasionally needed to achieve such an intake, and its documentation. In practice, however, relevant investigations are often undertaken during this trial of feeding. These may include:

- FBC, ESR, differential, and red cell indices – assess anaemia, neutropenia (Schwachman–Diamond syndrome).
- Ferritin; total iron-binding capacity – iron deficiency.
- Red cell folate – may be low in coeliac disease.
- U&E, creatinine, glucose and bicarbonate.
- Calcium, phosphate, alkaline phosphatase – may be abnormal in severe malabsorption or renal osteodystrophy.
- Liver function tests.
- Stool culture for bacteria, viruses, ova, cysts and parasites.
- Stool analysis for fat globules, faecal chymotrypsin and reducing substances with sugar chromatography.

- Urine for sugar and protein analysis, culture and pH measurement.
- Thyroid function tests.
- Immunoglobulins.
- Sweat test to exclude cystic fibrosis.
- Chest radiograph – for evidence of suppurative lung disease and, rarely, the metaphyseal dysostosis of Schwachman–Diamond syndrome.
- Other tests may be required, depending on the clinical setting, e.g. oesophageal pH monitoring to assess gastro-oesophageal reflux, jejunal biopsy, barium follow-through, HIV status, chromosomal analysis, etc.

For a minority of children who are failing to thrive, a specific organic cause can be identified and treated, but for the majority the approach needs to focus on the reasons why a particular child is being undernourished in the home setting, and should aim to modify these as appropriate. A multidisciplinary approach involving paediatrician, dietitian, speech therapist, social worker and clinical psychologist may be most valuable in difficult management problems.

Further reading

Marcovitch H. Failure to thrive. *BMJ,* 1994; **308:** 35–8.
Skuse D. Failure to thrive: current perspectives. *Current Paediatrics.* 1992; **2 (2):** 105–10.
Skuse D. Identification and management of problem eaters. *Archives of Disease in Childhood,* 1993; **69:** 604–8.

Related topics of interest

Behaviour (p. 35)
Child abuse (p. 59)
Chronic diarrhoea (p. 72)
Growth – short and tall stature (p. 149)
Nutrition (p. 249)

FLOPPY INFANT

Hypotonia in infancy may be due to a paralytic or a non-paralytic disorder. Paralytic conditions cause hypotonia with weakness and may affect the anterior horn cells, nerve fibres, neuromuscular junctions or muscles. Non-paralytic conditions cause hypotonia without significant weakness. The commonest cause of a floppy baby is perinatal asphyxia. The paralytic causes of floppiness are rare, and many are genetically determined, the commonest being spinal muscular atrophy.

Aetiology

1. *Paralytic*

- Spinal cord disorders. Trauma (e.g. birth injury), tumours.
- Anterior horn cell disease, e.g. spinal muscular atrophy, poliomyelitis.
- Peripheral neuropathy. Very rare in infancy.
- Neuromuscular disorder, e.g. neonatal or congenital myasthenia gravis.
- Congenital myopathy. Structural (e.g. central core disease, muscular dystrophy, myotonic dystrophy) or metabolic (glycogen and lipid storage disorders, periodic paralysis).

2. *Non-paralytic*

- Disorders affecting the CNS, e.g. birth asphyxia, hypotonic cerebral palsy, Down's syndrome, metabolic disorders (e.g. aminoacidurias).
- Connective tissue disorders, e.g. Ehlers–Danlos.
- Prader–Willi syndrome.
- Metabolic and endocrine disorders, e.g. hypercalcaemia, hypothyroidism.
- Benign congenital hypotonia.

Clinical features

Hypotonia at birth is usually due to severe hypoxia, but is occasionally due to drugs taken by the mother, e.g. diazepam. Severely hypotonic infants lie in the characteristic frog-like position, and on ventral suspension the four limbs hang down and the infants are unable to hold their head up. The head lags on pulling to the sitting position. There may be deformity of the chest and a 'see-saw' respiratory pattern. Infants with cerebral palsy may go through a hypotonic phase before becoming spastic. Benign congenital hypotonia is an idiopathic, non-progressive condition which tends to improve with age, and is a diagnosis of exclusion.

Investigation

- U&Es.
- Thyroid function.

- Serum calcium.
- Chromosomes.
- Investigations for inborn errors of metabolism.
- Brain imaging (e.g. CT or MRI scan). May be evidence of hypoxic damage, e.g. atrophy, cysts.
- Neurophysiology.
- Muscle biopsy.

Spinal muscular atrophy (SMA)

There are three main types of SMA which are due to progressive death and loss of anterior horn cells. The cause is unknown. Type I (Werdnig–Hoffmann disease) presents in the first few weeks of life with weakness and wasting of the muscles. Fasciculation is seen, particularly in the tongue. There may have been decreased fetal movements and the infant may be profoundly floppy at birth. The facial and bulbar muscles are unaffected so the infant has an alert expression and is able to swallow normally. It is an autosomal recessive condition with a prevalence of 1 in 20 000 births. It is rapidly progressive, with the majority dying of respiratory failure within 18 months. Type II is also autosomal recessive and is a more chronic condition which usually presents between 3 and 12 months of age. It causes severe muscle wasting, contractures and scoliosis, and the majority of affected children die by the age of 10 years. Children with type III (juvenile) SMA develop normally until about 2 years of age, when they develop limb girdle weakness, a waddling gait, and gradual loss of ability to walk. Progression of weakness is episodic and survival into adult life is usual.

Further reading

Roland EH. Muscle disorders in the newborn. In: Roberton NRC, ed. *Textbook of Neonatology*, 2nd edn. Edinburgh: Churchill Livingsone, 1992; 1103–14.

Related topics of interest

Inborn errors of metabolism (p. 191)
Muscle and neuromuscular disorders (p. 224)
Neonatal convulsions (p. 227)

GASTROENTERITIS

Acute gastroenteritis remains a significant paediatric problem, accounting for 5 million deaths annually in the developing world in the under-5 age group, where it often occurs against a background of malnutrition. In Western nations, mortality is now minimal, but gastroenteritis remains an important cause of morbidity, and a common reason for paediatric hospital admissions (second only to respiratory illness). The promotion of adequate fluid balance, good nutritional status and the avoidance of unnecessary drug therapies should be the priorities of all health workers.

Problems

- Dehydration.
- Metabolic disturbances.
- Oliguria/acute renal failure.
- Septicaemia.
- Protracted diarrhoea.
- Malnutrition.

Aetiology/pathogenesis

A pathogen may be identified in up to 80% of cases using modern techniques.

1. Viral

- Rotavirus (commonest in UK).
- Enteric adenovirus.
- Small round viruses (astro calici).
- Norwalk.

Rotavirus infection occurs in seasonal outbreaks (December to February), the virus attacking and killing the mature enterocyte, which is then shed from the small intestinal villus. The increased production of immature crypt-like cells with villus shortening then decreases the absorptive and disaccharidase activity of the brush border.

2. Bacterial

- *Campylobacter jejuni* (commonest invasive pathogen in UK).
- *Shigella* spp.
- *Salmonella* spp.
- *E. coli.*
- *Vibrio cholera.*
- *Yersinia enterocolitica.*

Various mechanisms may account for bacterial pathogenicity, and several mechanisms may be involved simultaneously.

- Mucosal invasion, e.g. *Campylobacter jejuni, Shigella* spp.
- Production of cytotoxins which alter mucosal surface properties, e.g. enteropathogenic *E. coli.*
- Production of enterotoxins which alter enterocyte salt and water balance, e.g. *Vibrio cholera*, enterotoxigenic *E. coli, Shigella* spp.
- Adherence to enterocyte surface, destroying the microvilli, e.g. enteropathogenic and enterohaemorrhagic *E. coli.*

3. Parasitic
- *Giardia lamblia.*
- *Cryptosporidium.*
- *Entamoeba histolytica.*

Again, several mechanisms may be involved, e.g. parasitic enterotoxin production, altered motility.

Clinical features

A prodromal illness is uncommon, but more likely in viral infection. Vomiting may precede the onset of watery diarrhoea by up to 48 hours. Abdominal cramps may be a feature. Blood and mucus in the stools, with associated fever, suggest an invasive organism. As fluid losses from the gastrointestinal tract proceed, dehydration occurs (see features below). Convulsions may occur from pyrexia or metabolic disturbances. Assessment of the abdomen is vital, as blood in the stool, with persistent vomiting, may be the presenting symptoms of an intussuception, for example.

Features of dehydration
- <5% (mild) — Not unwell, dry mucus membranes, thirsty.
- 5–10% (moderate) — Lethargic – sunken fontanelle, sunken eyes, reduced skin turgor, oliguria.
- >10% (severe) — Shocked – hypotension, peripheral shut-down.

Investigation

- Weight. Accurate weighing is critical for calculating fluid requirements and assessing the success of rehydration.
- U&E, creatinine, glucose. No biochemical tests are needed in mild cases.
- Urinary urea, creatinine, sodium. Consider these in incipient renal failure.
- Stool. Culture, electron microscopy, enzyme-linked

immunosorbent assay (ELISA). Pathogens can be identified, occasionally indicating the need for specific antimicrobial therapy
- Urine, blood, CSF. Consider culture if systemic sepsis is suspected.

Management

The aims should be:

1. Prevention and correction of dehydration. Oral rehydration solution (ORS) can be effective in 90% of moderately dehydrated patients. These solutions contain sodium, chloride and glucose in accurately defined ratios, and their success depends on the glucose- or amino acid-coupled sodium absorption across the enterocyte. Higher sodium concentrations are used in WHO ORS to approximate the sodium losses of cholera stools than the commercially available solutions in the UK, where stool sodium losses are lower, e.g. 90 mmol/l in WHO ORS, 60 mmol/l in Dioralyte (Rorer). If there is shock, coma, ileus or excessive stool loss, intravenous therapy is indicated.

If the patient is in shock, give 20 ml/kg colloid or 0.9% sodium chloride over 30 minutes.

Rehydrate, giving the fluid deficit as 0.18% sodium chloride, 4% dextrose (or 0.45% sodium chloride initially if serum sodium is 130 mmol/l or less). Add potassium chloride (3 mmol/kg/24 h) once urine is flowing. Maintenance fluid requirements in addition to ongoing fluid losses must also be given. For example a 15-kg child who is 5% dehydrated requires a ‘deficit volume’ = 5/100 x 15 x 1000 ml = 750 ml plus a ‘maintenance volume’ = 1250 ml, i.e. 100 ml/kg for the first 10 kg of body weight, 50 ml/kg for the next 10 kg of body weight and 20 ml/kg for weights above 20 kg. Therefore the total volume given in 24 hours = 2000 ml.

2. Promotion of adequate nutrition. Breast-feeding should continue uninterrupted throughout the illness. Traditional guidelines have recommended a slow reintroduction of formula milk after rehydration – ‘regrading’ (one-quarter, half, three-quarters, full-strength feeds over 12–72 hours), in an attempt to decrease the likelihood of protracted diarrhoea, but the incidence of lactose intolerance and cow's milk allergy appears to be declining, and routine ‘regrading’, certainly in the older infant, may no longer be indicated.

Indeed, there may be advantages in continuous feeding throughout the illness, or the use of a rice- or cereal-based ORS, particularly in the developing world.

3. Avoidance of metabolic disturbance. In moderate and severe cases, attention should be paid to plasma electrolytes. Hypernatraemia is now rare, due to the use of modified formula milks, but should be corrected by very slow, preferably oral, rehydration. Acidosis normally self-corrects with improved tissue perfusion.

4. Appropriate use of drugs. Antibiotics should not be used routinely, but may have a specific role for certain pathogens, e.g. metronidazole for giardiasis. Antidiarrhoeal agents should not be used – they do not reduce fluid losses and may have serious side-effects, e.g. paralytic ileus has been described with loperamide, respiratory depression with diphenoxylate.

Further reading

Turnberg L. Cellular basis of diarrhoea. *Journal of the Royal College of Physicians,* 1991; **25:** 53–62.

Walker-Smith JA. Management of infantile gastroenteritis. *Archives of Disease in Childhood,* 1990; **65:** 917–8.

Related topics of interest

Chronic diarrhoea (p. 72)
Nutrition (p. 249)

GASTROINTESTINAL HAEMORRHAGE

Gastrointestinal haemorrhage is always a worrying symptom, but a thorough clinical evaluation often reveals that the cause is benign and simply treated. The use of certain investigations may be necessary, especially for the 'silent bleeder', i.e. occult blood loss with no clinical clues as to its site of origin, and for those children in whom serious pathology exists.

Problems

- Haematemesis.
- Melaena.
- Bright-red rectal bleeding.
- Iron deficiency anaemia from occult blood loss.

Consideration should also be given to the 'mechanism' of bleeding, e.g. a generalized bleeding disorder (leukaemia, factor VIII deficiency), vasculitis (Henoch–Schönlein purpura, hereditary haemorrhagic telangiectasia) or artefactual bleeding (Munchausen syndrome by proxy, swallowed maternal blood from a cracked nipple).

Aetiology

Bleeding may occur from many different sites in the gastrointestinal tract. The age of the patient may be a good guide as to the likely cause.

1. Upper tract bleeding

- Oesophagitis.
- Hiatus hernia.
- Mallory–Weiss tear.
- Oesophageal varices (portal hypertension).
- Gastritis.
- Gastric and duodenal ulceration.
- Volvulus.
- Meckel's diverticulum.
- Mesenteric thrombosis.
- Perforated viscus.

2. Lower tract bleeding

- Polyps (Peutz–Jeger, familial adenosis polypi, juvenile colonic polyps).
- Duplications.
- Intussusception.
- Colitis (allergic, ulcerative, infective, Crohn's, necrotizing).
- Foreign body.
- Trauma (sexual abuse, insertion of rectal thermometer).

- Rectal prolapse.
- Anal fissure.
- Haemorrhoids.
- Angiodysplasia (very rare).

Consider also the drug history e.g. non-steroidal anti-inflammatory drugs (NSAIDs), steroids, salicylates.

Clinical assessment

In acute large bleeds, resuscitation should precede investigation. The history may reveal an infectious disease contact, trauma, associated symptoms (diarrhoea in ulcerative colitis, excessive vomiting in Mallory–Weiss, etc.) or a relevant family history, e.g. familial adenomatous polyposis coli. The colour and amount of blood should be noted and a stool examined by the clinician when possible. The stigmata of chronic liver disease, and those of Peutz–Jeger (perioral pigmentation) and hereditary telangiectasia may be evident on examination. Anal signs may be those of trauma or fissure. Careful abdominal assessment is mandatory. Gentle rectal examination, using a sweeping motion of the finger around the pelvis, may discover a polyp or a pelvic mass.

Investigation

- FBC. Anaemia, hypochromic normocytic picture of chronic blood loss, neutrophil leucocytosis, thrombocytopenia.
- Coagulation studies.
- Abdominal radiography. Of value in intussusception, ulcerative colitis (toxic dilatation), volvulus.
- Abdominal ultrasound.
- Barium studies. Polyps, intussusception, ulcerative colitis, oesophageal varices.
- Stool microscopy and culture.
- Apt's test (distinguishes maternal from neonatal blood).
- Endoscopic studies. Also facilitates biopsy but requires experienced operators and appropriate-sized endoscopes.
- Meckel's scan (technetium-99 pertechnetate).
- Angiography (rarely needed).

Management

Clearly this will depend on the identified cause, e.g. surgical intervention for bleeding Meckel's diverticulum. Single colonic polyps may be snared endoscopically. Antibiotics, anti-inflammatory drugs or an exclusion diet may be appropriate for colitis, depending on aetiology. Children with bleeding oesophageal varices should be stabilized before being transferred to a centre with appropriate expertise (see below). Rarely, if the bleeding source is unidentified, laparotomy and operative endoscopy is needed.

Management of bleeding oesphageal varices

- Matched whole blood or packed cells with fresh-frozen plasma should be transfused to achieve a normal blood pressure and adequate tissue perfusion. Overtransfusion may distend varices and worsen bleeding. Central venous pressure measurement may be useful.
- Regular aspiration of a wide-bore nasogastric tube will allow detection of bleeding and measurement of gastric pH, and will prevent further passage of blood into the intestine.
- Precipitation of encephalopathy should be avoided. Lactulose and neomycin reduce colonic pH and ammonia reabsorption while encouraging bowel emptying. Sedatives and diuretics should be avoided, and sepsis treated vigorously.
- H_2 blockers (cimetidine, ranitidine) or omeprazole, a proton pump inhibitor, should be used to reduce bleeding from gastric erosions. Standard doses should be reduced.
- Intravenous vasopressin causes constriction of the splanchnic arterial bed and may temporarily stop bleeding. Give 0.33 units/kg over 20 minutes, then this dose each hour by continuous infusion. It causes pallor and abdominal cramps.
- A Sengstaken–Blakemore tube may control bleeding by balloon tamponade using the oesophageal and/or gastric balloon. Malpositioning of the balloons may cause erosive damage and cardiac and respiratory complications.
- Endoscopy will identify the source of the bleeding (varices and/or gastric ulceration), but also provides the opportunity for endoscopic injection sclerotherapy.
- Rarely, surgical portosystemic shunts are needed, but there is a high risk of worsening liver function and hepatic encephalopathy.

Further reading

Raine PAM. Investigation of rectal bleeding. *Archives of Disease in Childhood,* 1991; **66:** 279–80.

Related topics of interest

Anaemia (p. 20)
Liver disease (p. 208)
Purpura and bruising (p. 278)

GLOMERULONEPHRITIS

Glomerulonephritis (GMN) is inflammation and proliferation of cells within the renal glomerulus which may occur as a primary condition or associated with a systemic disorder. The majority of cases are immunologically mediated. GMN may present with asymptomatic haematuria, acute nephritic syndrome or a mixed picture of nephritis and nephrosis.

Problems

- Haematuria.
- Hypertension.
- Renal failure.

Aetiology and pathology

In the majority of cases the damage is immunologically mediated, either by deposition of immune complexes within the glomerulus or by interaction of autoantibodies with the glomerular basement membrane. This leads to activation of the complement system and release of cytokines which attract other mediators of glomerular damage, such as neutrophils. Damage may be limited to the kidneys or be part of a systemic disorder. In a few cases the pathogenesis is unknown.

1. *Primary GMN*

- Immune complex GMN
  - Post-infectious acute nephritis (the majority are post-streptococcal).
  - Mesangial IgA nephropathy (Berger's disease).
  - Membranous and membranoproliferative GMN.
- Anti-glomerular basement membrane antibody-mediated GMN, e.g. Goodpasture's disease.
- Pathogenesis unknown.

2. *GMN associated with systemic disorders*

- Immunologically mediated.
  - Henoch–Schönlein purpura (microscopic haematuria in > 70%).
  - Systemic lupus erythematosus and other connective tissue disorders.
  - Vasculitic conditions, e.g. polyarteritis nodosa.
  - Infections, e.g. subacute bacterial endocarditis, shunt nephritis, malaria.
- Hereditary, e.g. Alport's syndrome (sensorineural deafness).
- Other, e.g. diabetes.

Light microscopic changes have led to the description of many categories of GMN, e.g. focal, diffuse, segmental, proliferative, membranous, sclerotic. Immune complexes may be identified by electron microscopy or by immunofluorescence microscopy.

Acute nephritic syndrome (acute post-infectious GMN)

This typically occurs 7–14 days after a group A β-haemolytic streptococcal throat infection but can follow other streptococcal infections (e.g. skin) or infections with other organisms, e.g. *Staphylococcus, Salmonella, Mycoplasma,* and viruses (coxsackie, echo, influenza). It can occur at any age but predominantly affects school-age children.

Clinical features

- Dark 'smoky' urine.
- Oliguria.
- Non-specific malaise.
- Hypertension (occasionally acute causing encephalopathy and seizures).
- Fluid retention (usually mild causing only a degree of oedema but if severe can result in heart failure).

Investigation

- Urinalysis. Gross haematuria, granular and red cell casts, variable proteinuria. If proteinuria is heavy, hypo-albuminaemia may occur and the patient will develop the clinical picture of the nephrotic syndrome.
- Plasma U&E, creatinine, calcium, phosphate, bicarbonate. Degree of renal failure may be minimal to severe.
- Evidence of recent streptococcal infection. Throat swab, blood for anti-streptolysin O (ASO) and anti-DNA antibody titres. ASO may not rise, particularly after skin infections.
- Complement screen and immunoglobulins. Low C3 and C4 during acute phase, return to normal within 8 weeks.
- Autoantibody screen, e.g. antinuclear factor positive in lupus nephritis.

Management

Acute GMN is usually mild and requires no specific treatment other than monitoring. There is no real evidence that bed rest is of benefit. Strict fluid balance with daily weighing will limit the risk of fluid overload. Dialysis is rarely needed. Antihypertensive treatment may be required if the blood pressure is not controlled by fluid restriction. Although it does not seem to alter the course of acute GMN, a 10-day course of penicillin is usually given to limit the spread of nephritogenic strains. Renal biopsy is not indicated unless renal function rapidly deteriorates after the acute phase or renal failure persists for several weeks. The child can be discharged home once the renal function is seen to be improving. Out-patient follow-up should continue until urinalysis, blood pressure and renal function are normal.

Outcome

Complete recovery occurs in over 95% of children with post-streptococcal GMN. Second attacks are rare. Renal function usually returns to normal within 10–14 days, but microscopic haematuria may persist for 1–2 years. In a few children the disease is fulminant with rapid progression to renal failure. In others there is slow deterioration of renal function over several months, leading to chronic or end-stage renal failure.

Further reading

MacDonel, Barakat AY. Glomerular disease. In: Baraket AY, ed. *Renal Disease in Children; Clinical Evaluation and Diagnosis*. New York: Springer Verlag, 1989; 171–84.

Related topics of interest

Acute renal failure (p. 10)
Chronic renal failure (p. 75)
Haematuria (p. 153)
Nephrotic syndrome (p. 243)

GROWTH – SHORT AND TALL STATURE

Growth assessment is fundamental to all areas of paediatrics. Poor growth may result from many disease processes and may be the earliest sign of endocrine abnormality. Growth from conception into infancy is almost entirely dependent on nutrition. During infancy growth is rapid; growth hormone (GH) receptors become detectable towards the end of the first year of life, and the slowly decelerating phase of growth during childhood is under GH control. During childhood, height increases by 5–7.5 cm/year. The pubertal growth spurt is controlled by both GH and sex steroids, and reaches a peak of between 7 and 12 cm/year. Pulsatile GH is secreted by the anterior pituitary, particularly at night. Secretion is stimulated by growth hormone-releasing hormone (GHRH) and inhibited by somatostatin. These are secreted by the hypothalamus under the influence of many factors, including neurotransmitters, glucose, amino acids, lipids, insulin-like growth factor 1 (IGF-1), sleep and temperature.

Growth assessment

- Anthropometric equipment for standing/sitting heights and infant length.
- Distance centile charts. Plotted against decimal age.
- Growth velocity charts. Growth rate in cm/year. Measure over as long a period as possible (preferably 1 year).
- Parental heights corrected for the sex of the child by adding 12.6 cm to mother's height for a boy or subtracting 12.6 cm from the father's height for a girl. The arithmetic mean of the adjusted parental heights ± 9 cm gives expected adult height with 95% confidence limits.
- Pubertal staging.
- Weight.
- Head (occipitofrontal) circumference.
- Bone age. Left wrist radiograph.
- Skinfold thickness. Useful in nutritional problems.

Short stature

Aetiology

1. Normal growth velocity

- Constitutional short stature. Three per cent of the normal population will be below the third centile.
- Low birth weight. Silver–Russell syndrome.
- Physiological growth delay. Delayed bone age, normal height prognosis.

2. Low growth velocity

- Chronic illness, e.g. renal, cardiovascular, respiratory disease.
- Malabsorption, e.g. Crohn's disease, coeliac disease.
- Undernutrition.

- Psychosocial deprivation.
- Syndromes, e.g. Turner's, Prader–Willi.
- Skeletal dysplasias. Disproportionate body/limbs.
- Endocrine disease. GH insufficiency, panhypopituitarism, hypothyroidism, Cushing's syndrome (usually iatrogenic).

Growth hormone insufficiency

1. *Congenital*
- Idiopathic GHRH deficiency.
- GH gene deletion (very rare).
- Developmental abnormalities, e.g. midline brain defects.

2. *Acquired*
- Tumours of hypothalamus or pituitary.
- Cranial irradiation.
- Transient insufficiency. Low sex hormone concentrations, psychosocial deprivation.

Investigation

1. *Growth assessment* (see above). Serial height measurements should be made over a minimum period of 3 months. Growth of less than 4 cm/year or a child growing between this rate and the 25th velocity centile for 2 years requires investigation. Ideally, poor growth should be detected before the child falls below the third centile.

2. *History and clinical examination* may identify evidence of chronic illness, malabsorption or malnutrition prompting appropriate investigations. Hypothyroidism should be excluded. Features of syndromes may be present, and Turner's syndrome should be excluded in girls. Children with GH insufficiency are said to have a characteristic appearance (truncal obesity, central crowding of facial features, immature facial appearance).

3. *Growth velocity.* If no disease is identified and the growth velocity is normal, it is doubtful whether any treatment can make a substantial difference to the final height. If the growth velocity is low, investigation depends on the age of the child. Under 1 year the cause is likely to be nutritional in origin. In childhood, GH insufficiency should be suspected. During normal puberty an increase in sex steroid secretion is associated with increased amplitude of pulsatile GH secretion. In a child of pubertal age with no signs of puberty, a low growth velocity nearly always indicates delayed onset of puberty. Treatment with sex steroids will stimulate both puberty and growth. If there are

signs of puberty but there is no pubertal growth spurt, GH insufficiency is likely.

4. *Tests of GH secretion.* Random GH levels are rarely of use owing to the pulsatile pattern of secretion. Measurement of 24-hour urinary GH requires an ultrasensitive GH assay to detect the low levels associated with GH insufficiency. GH provocation tests (e.g. insulin, clonidine or GHRH stimulation) may be indicated, although their use and interpretation is controversial (see Further reading). Children of pubertal age need priming with sex steroids prior to provocation tests. Twelve- or 24-hour profiles of GH secretion remain a research investigation at present, but abnormalities of pulsatility may be important. IGF-1 is GH dependent and levels are low in GH insufficiency. Levels are also low in young children and are affected by liver disease and hypothyroidism.

Management

Pituitary-derived GH was available in limited supplies from 1958 to 1985 but was withdrawn because of the risks of Creutzfeldt–Jakob disease. Biosynthetic GH is now available for use in GH insufficiency and is given by daily subcutaneous injection. It has been used in many other conditions, e.g. renal failure, Turner's syndrome and skeletal dysplasias, with some increase in growth velocity, but the effect on the final height of these children is still unclear.

Tall stature

Aetiology

- Constitutional, familial.
- True precocious puberty.
- Thyrotoxicosis.
- Excessive androgen secretion (adrenal tumour, congenital adrenal hyperplasia).
- GH excess (gigantism), e.g. pituitary adenoma.
- Others, e.g. Marfan's syndrome, homocystinuria, chromosomal abnormalities (e.g. Klinefelter's syndrome (XXY)).

Investigation

Thyrotoxicosis and precocious puberty (true and false) should be excluded by history, examination and appropriate investigations. It is often difficult to distinguish between a normal tall child and a child with gigantism, but an abnormal GH profile or a raised IGF-1 in the presence of a growth velocity above the 90th centile, or a greater than expected height for the family, should be indications to consider pituitary imaging. GH suppression tests used in the diagnosis of acromegaly in adults are rarely useful in children.

Management

Prevention of excessive height is better than trying to lose height already gained. Children with excessive predicted adult heights are currently treated with sex steroids to induce early puberty, but the long-term effects of early exposure, particularly to oestrogen, are unclear. Reduction of endogenous GH secretion by anticholinergic agents or somatostatin are areas of research interest.

Further reading

Brook CGD, Hindmarsh PC. Tests for growth hormone secretion. *Archives of Disease in Childhood,* 1991; **66:** 85–7.

Brook CGD. Who's for growth hormone? *BMJ,* 1992; **304:** 131–2.

Brook CGD. Growth. In: Brook CGD. *A Guide to the Practice of Paediatric Endocrinology.* Cambridge University Press, 1993; 18–54.

Shah A, Stanhope R, Matthew. Hazards of pharmacological tests of growth hormone secretion in childhood. *BMJ,* 1992; **304:** 173–4.

Related topics of interest

Chromosomal abnormalities (p. 68)
Puberty – precocious and delayed (p. 274)
Thyroid disorders (p. 316)

HAEMATURIA

Haematuria may occur as an isolated symptom or as part of a systemic disorder. It may be visible to the naked eye (frank or macroscopic haematuria) or be detected only on microscopic analysis of the urine. The source of the blood may be anywhere from glomerulus to urethra, but most cases in childhood are due to primary glomerular disease.

Causes

- Urinary tract infection (UTI).
- Glomerulonephritis (post-streptococcal, Henoch–Schönlein purpura, familial).
- IgA nephropathy.
- Acute haemorrhagic cystitis. Viral (adenovirus 11 and 21), drugs (cyclophosphamide).
- Calculus.
- Trauma.
- Exercise-induced. Usually after severe exercise and resolves within 48 hours.
- Tumours. Wilms' tumour (uncommon presentation), bladder tumours (rare in children).
- Subacute bacterial endocarditis.
- Infections, e.g. tuberculosis, schistosomiasis.
- Coagulopathies.
- Sickle cell disease. Sickling within the renal medulla leads to local papillary infarcts.
- Renal vein thrombosis. Gross haematuria and palpable renal mass in newborn infant.
- Factitious haematuria. As part of the Munchausen by proxy spectrum.

Clinical features

The urine is usually pink or brown in colour because of the presence of the oxidized haem pigment. In post-streptococcal GMN the urine is often described as smoky. If the urine is bright red with or without clots then a lower urinary tract source should be suspected. Haematuria may be an isolated finding or be associated with symptoms of a systemic disorder, e.g. Henoch–Schönlein purpura (rash, joint pains). Hypertension and oliguria are features of acute GMN. Frequency and dysuria suggest a UTI which may be accompanied by microscopic haematuria, but presentation with frank haematuria is rare. Loin pain or renal colic suggests the presence of a calculus. SBE, sickle cell disease and coagulation disorders are infrequent causes, but haematuria is rarely the presenting symptom.

Other causes of dark urine should be excluded:

- Bile pigments.
- Haemoglobinuria, myoglobinuria.
- Foods, e.g. beetroot.
- Drugs, e.g. rifampicin.
- Blood from other sources, e.g. menstrual, perianal.
- Urate crystals may appear pink in the nappy of young infants.

Investigation

- Urinary dipsticks are very sensitive and are therefore not very reliable. They are also positive for myoglobin and free haemoglobin.
- Urine microscopy should always be performed to confirm the presence of red cells (a fresh specimen is important since red cells lyse on standing). The presence of red cell casts indicates an intrarenal cause (either glomerular or tubular) and the red cells may be misshapen owing to distortion as they pass through the glomerular capillary wall. Glomerular casts indicate a glomerular cause. Pyuria and bacteriuria point to an infective cause which should be confirmed by culture. Microscopy may identify the ova of schistosoma if there has been recent travel to the tropics.
- Other tests depend on the history, examination and urinalysis:

  Urine culture will confirm a bacterial infection and lead to further appropriate investigations.

  Full blood count, film and coagulation tests. To exclude coagulopathies and sickle cell disease.

  Investigations for glomerulonephritis.

  Radiology. An abdominal US scan may identify a renal mass, evidence of obstruction or a renal tract calculus. Calculi may be seen on a plain abdominal radiography but an intravenous urogram may be needed to confirm the site of obstruction. Bladder causes of haematuria are very rare in childhood and cystoscopy is seldom indicated.

IgA nephropathy

This is the commonest cause of recurrent asymptomatic haematuria and was first described by Berger in 1968. It presents in later childhood or early adult life with persistent microscopic haematuria and episodes of macroscopic haematuria associated with upper respiratory tract infections. It was originally regarded as a benign disease but it is now recognized to have a slowly progressive course in some cases, leading to hypertension, proteinuria and eventual chronic renal failure. Occasionally the disease is fulminant, with rapid progression to renal failure. The aetiology is unclear but diffuse mesangial deposits of IgA are seen on immunofluorescent microscopy. It is more common in southern Europe, South-East Asia and Japan, and it is twice as common in males.

Calculi and nephrocalcinosis

Urolithiasis (calculi in the renal tract) and nephrocalcinosis (deposition of calcium in the renal parenchyma) are uncommon in children. Calculi are associated with stasis (e.g. neuropathic bladder), UTIs, idiopathic hypercalciuria (normal serum calcium) and cystinuria. Nephrocalcinosis is associated with renal tubular acidosis, particularly the distal type.

Further reading

White RHR. The investigation of haematuria. *Archives of Disease in Childhood,* 1989; **64:** 159–65.

Related topics of interest

Glomerulonephritis (p. 146)
Urinary tract infection (p. 322)

HAEMOLYTIC URAEMIC SYNDROME

Haemolytic uraemic syndrome (HUS) is an uncommon but serious condition with an incidence of ~ 75 cases per year in the UK. The incidence has risen abruptly since the early 1980s, but this may be because of improved reporting. HUS is characterized by microangiopathic anaemia, thrombocytopenia and acute renal failure. Over 50% of patients require dialysis for acute renal failure and the mortality in the acute phase is 10–15%. The majority of cases are associated with a prodromal diarrhoeal illness (D+), but rare cases are idiopathic, inherited or drug associated (D–).

Problems

- Haemolytic anaemia.
- Thrombocytopenia.
- Renal failure.
- Hypertension.
- Neurological impairment.

Aetiology and pathology

1. Typical HUS (post-diarrhoeal, epidemic, D+) is associated with infections, especially verotoxin-producing *E. coli*, but is also described following viral infections (e.g. coxsackie), shigella dysentery and *Streptococcus pneumoniae*. The renal damage is primarily glomerular with a thrombotic microangiopathy causing multisystem disease. Endothelial injury may be mediated by activated neutrophils, or by the direct effect of endotoxin or verotoxin.

2. Atypical HUS (sporadic, D–) is rare and may be idiopathic, inherited or associated with drugs such as mitomycin C, crack cocaine, quinine. The renal damage is arteriolar with intimal and subintimal oedema, necrosis and proliferation.

Clinical features

1. Typical HUS (D+). The majority of cases of HUS are D+. The diagnosis should be considered in any child who develops pallor, oliguria or haematuria following a diarrhoeal illness, particularly if the diarrhoea is bloody. It occurs in summer epidemics, typically in infants and young children, and is equally common in boys and girls. Bruising and petechiae occur secondary to thrombocytopenia, and jaundice may develop. Hypertension, if present, is usually mild. CNS involvement (e.g. convulsions) is rare and the long-term prognosis is good (90% survival). Recurrences are unusual.

2. Atypical HUS (D–). This is uncommon and tends to occur in older children. There is no seasonal variation in incidence and there is usually no prodromal diarrhoea.

Hypertension is severe, CNS involvement is common and the prognosis is poor. It can recur in transplanted kidneys.

Investigation

- Full blood count and film. Thrombocytopenia ($< 100 \times 10^9/l$), microangiopathic haemolytic anaemia (Hb < 10g/dl, distorted and fragmented red cells), neutrophilia (poor prognostic factor if greater than $20 \times 10^9/l$).
- Fibrin degradation products (FDPs). Raised.
- Urea, creatinine, and electrolytes. Urea typically > 18.
- Stool culture. *E. coli* serotype 0157:H7 is the commonest verotoxin-producing *E. coli* (VTEC) associated with D+ HUS. It may be isolated for many weeks, but the earlier the stool collection the higher the chance of isolation.
- Elastase. Produced by activated neutrophils (higher levels are associated with worse outcome).

Management

1. Anaemia should be corrected with repeated small transfusions of packed cells.

2. Fluid and electrolyte balance. Record strict input and output charts with fluid restriction and daily weights. Dialysis is implemented early for fluid overload and a rising urea or potassium. Peritoneal dialysis is the method of choice.

3. Hypertension should be treated vigorously with antihypertensive agents, fluid restriction and dialysis.

4. Complications include respiratory failure and pancreatitis.

5. Specific therapies. Many have been tried, including heparin infusions, antiplatelet agents such as dipyridamole and infusions of fresh-frozen plasma but their benefit is unproven. Prostaglandin I_2 activity has been noted to be low in the acute phase, which may be the result of impaired synthesis or increased degradation. Interest currently surrounds the use of PGI_2 infusions in the acute phase.

Outcome

In a national study carried out in the UK of children diagnosed as having HUS between 1985 and 1988, Milford *et al.* (1990) identified 298 children. Ninety-five per cent of these were D+, and at follow-up 13% had either died or had significant renal morbidity (hypertension, chronic or end-stage renal failure). Of the children with D– HUS, 79% were dead, hypertensive or in renal failure. Other studies have shown significant residual nephropathy in as many as one-third of patients 5–10 years after the acute episode. These abnormalities of renal function may be subtle and indicate only abnormal renal fuctional reserve, but this may have long-term implications. The follow-up of these children should at least include regular monitoring of blood pressure, serum creatinine and urinalysis for protein.

Further reading

Milford DV, Taylor CM, Guttridge B, Smith HR, Scotland SM, Gross RJ, Hall SM, Rowe B, Kleanthous H. Haemolytic uraemic syndrome in the British Isles 1985–8: association with verotoxin producing *Escherichia coli*. 1. Clinical and epidemiological aspects. 2. Microbiological aspects. *Archives of Disease in Childhood,* 1990; **65:** 716–21, 722–7.

Related topics of interest

Acute renal failure (p. 10)
Chronic renal failure (p. 75)

HAEMOPHILIA AND CHRISTMAS DISEASE

Haemophilia A (factor VIII deficiency) is a sex-linked recessive condition which occurs in 1 in 20 000 males. A family history is obtained in over 50% of cases. Christmas disease (factor IX deficiency = haemophilia B) is also sex-linked recessive but is much less common. These coagulopathies may present in infancy with excessive superficial bruising, or bleeding following circumcision. Once the child is mobile, haemarthroses are common and bleeding post tonsillectomy may be life-threatening. Bleeding from cuts or from mucosal surfaces (e.g. epistaxis) is less common, but haematuria does occur. Female carriers are asymptomatic but can be detected by reduced factor VIII or IX activity.

Problems

- Haemarthroses.
- Muscle haematomas.
- Joint deformity and ankylosis.
- Elective surgery.
- Risk of viral infections from blood products.
- Family and social disruption.

Investigation

- Prolonged partial thromboplastin test (PTT).
- Normal prothrombin time (PT), thromboplastin time (TT) and fibrinogen.
- Bleeding time normal.
- Liver function tests.
- Clotting factor assays. In haemophilia A, levels of factor VIII coagulation activity (VIIIc) produced by the liver are low. Levels of factor VIII-related antigen (VIII RAG) produced by endothelial cells and platelets are normal. In Christmas disease, factor IX levels are low and some patients produce an abnormal factor IX (IXCAg). Assays of other vitamin K-dependent factors (II, VII, X) are sometimes necessary to exclude a hepatic disorder.
- Gene analysis is now available, making antenatal diagnosis possible.

Management

Christmas disease tends to have milder clinical manifestations than haemophilia A, but the general management principles are the same. The severity of haemophilia is linked to the degree of clotting factor deficiency: < 1% of normal causes severe problems, 1–5% causes moderate problems and 5–20% only mild symptoms. Factor VIII or factor IX concentrates are now used instead of cryoprecipitate in the treatment of acute bleeds. The amount of concentrate required depends on the desired rise in factor VIII or IX:

$$\text{Units of factor VIII} = \frac{\text{weight (kg)} \times \text{desired rise in factor VIII (\%)}}{1.5}$$

Following a head injury or severe bleed a level of over 60% is aimed for. Bleeds in the throat are particularly serious as they may cause respiratory embarrassment. Following a haemarthrosis or soft-tissue injury levels of 30–50% should be achieved. Haemarthroses cause pain and, if severe, the joint should be splinted to maintain a good position and to relieve pain. Opiate analgesia may be needed, but intramuscular injections should be avoided. Passive exercises and later active exercises are a vital part of recovery and prevention of joint deformity and ankylosis. Factor VIII or IX inhibitors can develop, which makes treatment much more difficult.

Tranexamic acid is an antifibrinolytic agent which is useful in controlling mucosal bleeds but is not useful for other bleeds and is contraindicated in haematuria.

Mild to moderate haemophilia may respond to desmopressin (DDAVP) which can be infused to cover minor procedures such as dental extraction.

Haemophilia and HIV infection

The use of multidonor cryoprecipitate and lyophilized factor VIII or IX concentrates, before the recognition of the HIV virus, has led to the infection of a significant number of haemophiliacs. In the UK in 1990, 17% of the 863 haemophiliacs under the age of 14 years and 42% of the 424 haemophiliacs between 15 and 19 years were HIV positive. Twenty of these children and adolescents had developed AIDS and there had been nine AIDS-related deaths. However, no child under 5 years was HIV positive, and as a result of exclusion of high-risk donors, HIV antibody testing and the treatment of clotting factor concentrates to remove viruses, no further cases of infection have been reported since 1986.

Haemophiliacs have also been at risk of chronic liver disease from transfer of hepatitis B and C in blood products. They should be immunized against hepatitis B, and it is hoped that the treatment methods used to eliminate HIV from clotting factors will also eliminate other viruses such as hepatitis C.

Further reading

Jones P. HIV infection and haemophilia. *Archives of Disease in Childhood,* 1991; **65:** 364–8.

Related topics of interest

HIV infection (p. 171)
Purpura and bruising (p. 278)

HEADACHE

Headache is a frequent symptom in childhood, and is usually benign and short-lasting. Less frequently it may be a symptom of serious underlying disease such as an intracranial neoplasm or encephalitis. Pain receptors are not present within the brain itself, but are found in the blood vessels at the base of the brain, in the meninges, the paranasal sinuses, teeth, eyes and the muscles of the head, face and neck.

Aetiology

1. Acute headache

- Acute febrile illness.
- Localized infection within the head, e.g. otitis media, sinusitis.
- Acute central nervous system infection, e.g. encephalitis, meningitis.
- Trauma – accidental or non-accidental.
- Hypertension.
- Intracranial haemorrhage, e.g. from a vascular malformation or ruptured aneurysm.
- Physical factors, e.g. hunger, noise.

2. Recurrent headache

- Migraine.
- Raised intracranial pressure, e.g. from neoplasm, hydrocephalus, benign intracranial hypertension.
- Tension.
- Depression or anxiety.
- Epilepsy.
- Metabolic problems, e.g. hypoglycaemia.
- Refractive errors.

History

If the headache is acute in onset, the history must elicit any symptoms of meningitis (fever, vomiting, photophobia), a history of recent trauma, or of other acute infection, e.g. otitis media. Paroxysmal headaches with nausea and a preceding aura are suggestive of migraine, and a positive family history is found in up to 90% of cases. Recurrent headaches which are present on morning waking, are accompanied by vomiting or which worsen with straining or coughing are suspicious of an intracerebral mass lesion (displacing or dilating intracranial vessels). Enquiry should be made about the child's gait and any recent personality changes. The social history should enquire about any obvious causes of stress.

Examination

In acute headache, the child's temperature and consciousness level should be recorded, and the child should be examined for neck stiffness. An assessment should be made of the child's growth, including head circumference. Examination of the skin may reveal stigmata of the neurocutaneous syndromes, which are associated with intracranial neoplasms. A full neurological examination should be performed, with particular attention to gait, fundoscopy, visual fields and visual acuity. Blood pressure should be measured in all cases. The cranium should be auscultated for bruits, occasionally heard with arteriovenous malformations. There may be evidence of localized pathology in the head and neck, e.g. teeth, sinuses.

Investigation

In some cases, repeated clinical assessment only is required. Selected investigations may include:

- FBC, ESR.
- U&E, glucose.
- Blood culture. In acute febrile illness.
- Skull radiographs exclude fracture. Features of raised intracranial pressure (suture diastasis, increased digital markings), suprasellar calcification (found in craniopharyngioma) may be found, but CT or MRI will be indicated if there is clinical suspicion of raised intracranial pressure.
- Sinus radiography is occasionally helpful in the older child, but difficult to interpret in the young child, as the sinuses are poorly developed.
- CT or MRI scan excludes hydrocephalus or a space-occupying lesion. MRI is the investigation of choice, especially to exclude posterior pathology.
- Cerebral angiography is occasionally needed to demonstrate local blood vessel anatomy prior to surgery.
- EEG. Slow-wave changes in encephalitis, epileptic focus.
- Lumbar puncture may include microscopy, culture and sensitivity, cytospin (if malignancy suspected), virology or a search for acid–alcohol-fast bacilli if tuberculous meningitis is suspected. Lumbar puncture is contraindicated if a mass lesion is suspected, or if there is raised intracranial pressure.

Management

Management depends entirely on the cause of the headache, e.g. appropriate antimicrobial treatment of CNS infection, with management of raised intracranial pressure, neurosurgical intervention for hydrocephalus, arteriovenous malformation, etc. Migraine is normally characterized by nausea, unilateral paroxysmal headache and focal neurological aura. Management may be both symptomatic (paracetamol, antiemetic drugs) and prophylactic (regular sleep and meals, avoidance of known food triggers, beta-blockers or pizotifen – an antihistamine and serotonin antagonist). Recurrent tension headaches in the absence of organic pathology need a combined approach of strong reassurance and strategies to reduce stress.

Further reading

Brett EM. Vascular disorders including migraine. In: Brett EM, ed. *Paediatric Neurology,* 2nd edn. Edinburgh: Churchill Livingstone, 1991; 556–650.

Related topics of interest

Hydrocephalus (p. 174)
Hypertension (p. 177)
Meningitis and encephalitis (p. 220)
Neurocutaneous disorders (p. 246)

HEARING AND SPEECH

Language development may be affected by environment, intelligence and genetic factors, by the auditory input received and by articulatory processes. One in 500 children has moderate to severe deafness, requiring a hearing aid.

Important milestones of hearing and language development are:

- *4–6 weeks:* Smiles, then vocalizes in response to sounds.
- *3–4 months:* Turns to sounds on a level with the ear.
- *7–8 months:* Produces a few consonants – ba, da – and imitates sounds of two syllables.
- *12 months:* Two to three words used with meaning.
- *15 months:* Jargon speech and many words.
- *21–24 months:* Puts 2–3 words together into a sentence spontaneously.
- *3 years:* Asks questions and understands much of adult speech. Vocabulary of 200 words, including pronouns.

Aetiology of deafness

1. *Perinatal.* Several factors may coexist, e.g. hyperbilirubinaemia, drugs (frusemide, gentamicin), neonatal meningitis, hypoxic ischaemic encephalopathy.

2. *Genetic.* Numerous syndromes exist, e.g. Treacher Collins, Alport, Usher, Waardenburg. Isolated deafness may also be genetically determined.

3. *Intrauterine disease.* Congenital infection with CMV, rubella, toxoplasma, syphilis and herpes simplex may cause deafness.

4. *Metabolic.* Mucopolysaccharidosis, (e.g. Hunter's syndrome) and Pendred's syndrome (hypothyroidism, autosomal recessive inheritance), may cause deafness.

5. *Acquired.* Acquired causes include trauma, meningitis and chronic secretory otitis media.

Aetiology of speech delay

- Hearing defects.
- Familial and genetic factors.
- Global developmental delay.
- Psychosis.
- Dysarthria.
- Dysrhythmia.

Clinical assessment

A detailed family history of language development and deafness is needed. The history of the pregnancy, delivery and post-natal period should be noted; any significant illnesses or trauma since birth may also be relevant. Enquiry should be made about the child's preverbal language development and current opportunities for talking with adults and peers. Maternal concerns about the child's hearing should always be taken seriously. Relevant abnormal behaviour, e.g. excessive gesturing, may be reported. A full clinical examination should search for features of syndromes associated with deafness and should include growth measurements and a formal assessment of development.

Hearing assessment

The test must be chosen according to the child's age and intelligence. The child surveillance programme should include enquiry into parental concerns, familial deafness and the child's startle reflex at 6–8 weeks of age, and a distraction hearing test at the age of 7–9 months. Surveillance between 18 months and 4 years is more difficult, relying on assessment of language, with audiology referral if there are doubts.

Methods for testing the neonate with risk factors for deafness include the auditory response cradle (ARC), brain-stem evoked response tests (BSER) and programmable otoacoustic emission by stimulation (POEMS).

The distraction test uses frequency-specific stimulli, e.g. the Manchester rattle, and may be used for infants from 7 months old.

Speech tests ('show me the fish') such as the Kendall or McCormick toy tests may be useful for the child below the age of 4 years. Performance tests, requesting a specific response every time a sound is heard, may also be of value.

From 4 years, pure-tone audiometry with headphones can be used, and this method forms the basis for screening hearing at school entry, 'the sweep'.

Other useful investigations are electrocochleography (ECOG) and brain-stem evoked responses – both may be useful if global developmental delay impairs responses in the distraction or performance tests. Tympanometry may be of value in assessing middle ear problems.

Language assessment

Language tests must include assessment of comprehension, expression, articulation and symbolic language, and are best performed by a speech therapist. The Reynell developmental language scales include assessment of both comprehension and expression. The Renfrew scheme tests expressive language (naming multiple pictures or telling the story of a picture). The Edinburgh articulation test concentrates on production of consonants. Symbolic language may be tested by the miniature toys test.

Investigation

For 60–80% of deaf children there is an identifiable cause. Investigation is important for genetic counselling and for management of associated problems.

- Serology. Rubella-specific IgM may be present in the neonate, or rubella IgG beyond 6 months of age.
- Urine collection. Test for blood and protein (Alport's syndrome), mucopolysaccharides, viral culture (CMV, rubella).
- Chromosomal analysis.
- Thyroid function (Pendred's syndrome). A perchlorate test should be included.
- ECG. A long QT segment may be present in the syndrome of Jervell and Lange-Nielsen.
- Radiology. Skull radiographs, cranial CT or MRI scan to investigate craniofacial abnormalities and bone dysplasias.
- Ophthalmic assessment. Cataracts and retinopathy in congenital rubella syndrome, heterochromia irides in Waardenburg's syndrome.

Management

Speech therapy and intervention programmes may be appropriate for the child with language disorders, with electronic devices and signing language for the more severely affected. The deaf child may need an amplification hearing aid, and may be helped by learning to lip read or by the use of British sign language, finger spelling or Makaton. Genetic counselling should be offered where appropriate. Appropriate provision should be made for the education of children with communication difficulties, either by adaptation within a normal school or by placement at a special school.

Further reading

Fonseca SJ, Patton MA. Clinical assessment of deafness. *Current Paediatrics,* 1992 **2(4):** 229–32.

Hall DMB, ed. *Health for all Children. A Programme for Child Health Surveillance.* Oxford: Oxford University Press, 1991.

Related topics of interest

Developmental assessment (p. 111)
Developmental delay and regression (p. 115)
Prematurity (p. 271)
Vision (p. 325)

HEART FAILURE

Inability of the heart to maintain cardiac output in childhood is usually associated with congenital heart disease which causes either pressure or volume overload. Signs and symptoms almost always develop in the first year of life, with biventricular failure resulting in pulmonary venous and hepatic congestion. Occasionally there is a non-cardiac cause such as fluid overload or anaemia. Heart failure is uncommon in older children. Management is aimed at removing treatable causes, reducing sodium and water load, improving myocardial contractility and general supportive measures such as maintaining good nutrition.

Problems

- Breathlessness.
- Sweating.
- Failure to thrive.
- Shock.

Aetiology

1. Presenting in the first week of life

- Severe left ventricular outflow tract obstruction, e.g. interrupted aortic arch, severe coarctation of the aorta, critical aortic stenosis.
- Hypoplastic left heart.
- Endocardial fibroelastosis.
- Myopathy (Pompe's, viral).
- Volume overload, e.g. cranial arteriovenous fistula.

2. Presenting in infancy

(a) Non-cardiac

- Fluid overload.
- Anaemia.
- Septicaemia.

(b) Cardiac

- Pressure overload, e.g. coarctation of the aorta.
- Volume overload, e.g. large VSD, PDA.
- Myocardial disease, e.g. myocarditis, metabolic defect, ischaemia (e.g. asphyxia).
- Arrhythmias, e.g. supraventricular tachycardia.
- Endocardial fibroelastosis.

3. Presenting in later childhood

- Systemic hypertension, e.g. renal disease.
- Pulmonary hypertension, (e.g. idiopathic), CHD with increased pulmonary blood flow or pulmonary venous obstruction, chronic lung disease (e.g. cystic fibrosis).
- Myocarditis (viral, rheumatic fever).
- Cardiomyopathy (including drug toxicity, e.g. anthracyclines).

Clinical features

Infants present with breathlessness and sweating particularly on exertion (feeding, crying). They take longer to complete a feed and failure to thrive is common. Examination reveals tachycardia, tachypnoea and hepatomegaly. There may be a murmur due to tricuspid/mitral valve incompetence, a congenital heart defect, or anaemia and a gallop rhythm may be heard. Hypotension occurs if heart failure is severe. Cyanosis develops if there is marked pulmonary oedema. Auscultation over the anterior fontanelle will reveal a bruit in the presence of an intracranial arteriovenous fistula. Basal crepitations and peripheral oedema do not usually occur in young children although there may be some facial puffiness. In neonates, if the heart failure has been present antenatally, there may be ascites, pleural and pericardial effusions and generalized oedema (hydrops fetalis). Older children exhibit clinical features of right or left heart failure as seen in adults (raised jugular venous pressure (JVP), hepatomegaly, peripheral oedema, etc.).

Investigation

- Chest radiography. Cardiomegaly, pulmonary plethora.
- ECG. Arrhythmias, evidence of ventricular strain or ischaemia.
- Echocardiography provides a measure of left ventricular function (e.g. fractional shortening) and identifies congenital heart defects.
- Other investigations to identify cause, e.g.

  Viral titres (especially coxsackie, echovirus).
  Thyroid function tests.
  Urine and plasma amino/organic acids (inborn errors of metabolism).
  Urinary glycosaminoglycans (mucopolysaccharidosis).
  Creatine kinase (myopathy).

Management

1. Oxygen. Supplemental oxygen is given via head box or face mask to maintain arterial saturations.

2. Diuretics remain the mainstay of treatment. Frusemide is the most widely used, with spironolactone added for its potassium-sparing properties. Fluid restriction may be of some benefit.

3. Treat anaemia. Small transfusions of packed cells with diuretic cover to prevent worsening of fluid overload.

4. Digoxin is used for its positive inotropic effect but controversy surrounds its value, particularly in situations of volume overload, e.g. VSD.

5. *Vasodilators,* e.g. captopril, are particularly useful in conditions in which there is poor left ventricular function, such as dilated cardiomyopathy.

6. *Nutritional support.* Heart failure leads to increased nutritional requirements owing to a raised metabolic rate. If the patient is very breathless, nasogastric feeds should be introduced, and calorie supplements should be considered at an early stage.

7. *Positive inotropes,* e.g. dopamine, are used in the intensive care setting to improve myocardial contractility.

8. *Surgery.* If heart failure remains poorly controlled or the child is severely failing to thrive, surgical correction of structural heart defects (e.g. large VSD) or pulmonary artery banding to reduce pulmonary blood flow may be necessary.

Further reading

Cardiac emergencies. In: Advanced Life Support Group. *Advanced Paediatric Life Support,* London: BMJ Publications, 1993: 101–4.

Related topics of interest

Arrhythmias (p. 26)
Congenital heart disease – acyanotic (p. 91)
Congenital heart disease – cyanotic (p. 95)
Shock (p. 291)

HIV INFECTION

AIDS (acquired immunodeficiency syndrome) results from infection with the human immunodeficiency virus (HIV). WHO estimated that 10 million adults and 1 million children would be infected by the end of 1992, with the majority living in sub-Saharan Africa. HIV is an enveloped retrovirus, containing genomic DNA, core proteins and reverse transcriptase. Adults and children differ in the manifestations of their disease.

Transmission

1. Parenteral. A major source of infection in the early 1980s was the contaminated factor VIII given to haemophilia sufferers. Since 1986, with routine heat treatment of coagulation factors, screening of blood and exclusion of 'high-risk' donors, these routes account for a diminishing proportion of infections. Adolescent intravenous drug abusers are at risk via this route.

2. Vertical. Transmission may occur prenatally, perinatally or post-natally. The European Collabarative Study (1992) has reported vertical transmission rates of 13–16% (up to 40% in Africa). Transmission efficiency increases with the presence of symptomatic HIV disease in the mother, premature delivery, the acquisition of HIV during pregnancy and breast-feeding. Breast-feeding confers an additional risk of transmission of 14%, and should be advised against, unless the estimated risks associated with bottle-feeding (the developing world) are felt to be unacceptably high.

3. Sexual. HIV infection by heterosexual or homosexual activity in teenagers may occur; cases resulting from child sexual abuse have been reported in USA.

Diagnosis

Early diagnosis is difficult, as the infected infant looks normal at birth, and maternal HIV antibody may persist for up to 18 months. However the potential use of antimicrobial and antiretroviral drugs means that immunological and viral tests (not all routinely available) are becoming more important, despite relative insensitivity in early infancy. The diagnosis is made by the presence of:

- Clinical features of AIDS.
- Persistent HIV antibody after 18 months.
- p24 antigen or virus in peripheral blood.

Polymerase chain reaction to detect and amplify HIV genetic material, *in vitro* antibody production (IVAP), high levels of IgG, IgM and IgA, and low CD4 numbers with reversed CD4:CD8 subsets may be additional helpful diagnostic indicators.

Clinical features

These reflect the viral effects on the developing immune system. The children presenting under 1 year with opportunistic infections or HIV encephalopathy have a poorer prognosis than those who present much later with bacterial infections or lymphoid interstitial pneumonitis (LIP). The 'latent' period between infection and symptoms is longer for parenterally infected patients than for those who are vertically infected.

The following are AIDS indicator diseases (1987 Centers for Disease Control):

1. Opportunistic infections. Pneumocystis carinii pneumonia (PCP) is the most common indicator disease in children. Features include wheeze, tachypnoea, hypoxaemia and a 'ground glass' appearance on chest radiographs. Diagnosis is confirmed by bronchoalveolar lavage with biopsy or immunofluorescence and monoclonal antibodies. Other infections include persistent oral or napkin candidiasis (common), *Cryptosporidium* and atypical *Mycobacterium avium intracellulare* infection and chronic infection with herpesviruses, e.g. CMV, varicella and herpes simplex.

2. HIV encephalopathy. This may present as developmental delay or loss of skills, ataxia, spastic diplegia, occasional seizures or acquired microcephaly. MRI or CT imaging may show mild brain atrophy with basal ganglial enhancement or white matter attenuation.

3. Lymphocytic interstitial pneumonitis. This affects 30–50% of vertically infected children and is a slowly progressive chronic lung disease characterized by bilateral reticulonodular infiltrates on chest radiographs (despite the absence of symptoms in the early stages). Later, features include cough, dyspnoea, clubbing and hypoxaemia and the picture may be confused further by superimposed PCP or lung infection with mycobacteria or other bacterial pathogens.

4. Bacterial infections. The common bacteria implicated in paediatric infection, especially the polysaccharide encapsulated organisms, are usually responsible for severe and recurrent infections, and are a significant cause of morbidity.

5. Severe failure to thrive. Most prevalent in developing countries, failure to thrive in infected patients may be worsened by malabsorption resulting from enteric infections, poor oral intake and the increased metabolic rate associated with severe systemic infection.

6. Malignancy. Kaposi's sarcoma is rare in children, but B- and T-cell lymphomas of liver, spleen, kidneys, bone marrow and CNS are seen.

Management

Close follow-up of potentially infected children at 3-monthly intervals is essential to monitor their physical well-being, growth, development and immunological status.

Prevention and early, aggressive therapy for infections are the mainstays of treatment. Routine childhood vaccinations should be given at the scheduled times. PCP prophylaxis can be given by cotrimoxazole three times weekly, with nebulized pentamadine as an option for the older child. Passive antibody protection by monthly intravenous infusion of immunoglobulin seems to reduce serious bacterial infection but not mortality. Antiretroviral treatment is available as AZT (3'-azido-3'-deoxythymidine) for children, and preliminary studies suggest that treated children show improvements in weight gain, well-being, and HIV encephalopathy. DDI (dideoxyinosine) and DDC (dideoxycytosine) are also antiretroviral agents under evaluation.

Further reading

Gibb D, Newell ML. HIV infection in children. *Archives of Disease in Childhood,* 1992; **67:** 138–41.

Jones P. HIV in children. *BMJ,* 1994; **308:** 425–6.

Related topics of interest

Haemophilia and Christmas disease (p. 159)

Immunodeficiency (p. 187)

HYDROCEPHALUS

Hydrocephalus is the presence of an increased amount of CSF, under increased pressure, with enlargement of the ventricular system. CSF is produced by the choroid plexuses of the ventricles, and passes from the lateral ventricles through the foramina of Monro into the third ventricle and through the aqueduct of Sylvius into the widening fourth ventricle, entering the subarachnoid space and the basal cisterns. It is then absorbed through the arachnoid villi into the cerebral venous sinuses. Hydrocephalus may therefore occur when there is overproduction of CSF, failure of reabsorption, or obstruction to its flow at any point. In non-communicating hydrocephalus, there is an obstruction within the ventricular system. In communicating hydrocephalus, the ventricular system is patent, but absorption of CSF by the arachnoid villi is prevented.

Arrested hydrocephalus implies that CSF pressure has returned to normal, with no pressure gradient between the ventricles and the brain parenchyma.

Aetiology

1. Congenital. Three main congenital malformations occur. Aqueduct stenosis consists of thickening in the tissues surrounding the aqueduct, leading to compression and distortion of the canal. The Arnold–Chiari malformation of the cerebellum involves a blockage to the outflow of the fourth ventricle by elongation of the medulla and downward displacement of the cerebellar tonsils. Myelomeningocele is commonly associated. The Dandy–Walker syndrome consists of cerebellar hypoplasia and obstruction to outflow foramina of the fourth ventricle. Other congenital anomalies presenting with hydrocephalus include an aneurysm of the vein of Galen (which may occlude the aqueduct), arachnoidal cysts and bony deformities of the base of the skull (e.g. achondroplasia) which may obstruct CSF circulation.

2. Neoplasia. Neoplastic lesions may cause obstruction to CSF circulation at various places. Papilloma of the choroid plexus is a rare cause of hydrocephalus, producing excess CSF.

3. Infection. Adhesions may obstruct the pathways of CSF flow after bacterial or tuberculous meningitis or following intrauterine infection, e.g. with *Toxoplasma*.

Clinical features

During infancy, presenting features include an enlarging head circumference (measurements must be serially plotted on growth charts), with an open large anterior fontanelle under increased tension. Vomiting, drowsiness, irritability and delayed motor milestones may occur. There may be dilated scalp veins, separation of the cranial sutures on palpation and a 'cracked pot' note when the skull is lightly percussed. The skull may transilluminate with a powerful light source. Close examination of the eyes may reveal the 'setting sun' sign, convergent squint with a sixth nerve palsy or restriction of upward gaze, but papilloedema is uncommon in infants because of the ability of the skull to expand.

Investigation

The diagnosis of hydrocephalus must be confirmed by further investigation, but the underlying cause must also be established.

- Skull radiograph. This is of limited value, often showing the non-specific signs of raised intracranial pressure (thin skull vault with suture diastasis), but occasionally diagnostic features may be found, e.g. suprasellar calcification in craniopharyngioma.
- CT scan. This confirms the hydrocephalus and, by demonstrating which ventricles are dilated, and showing space-occupying lesions, it may reveal the aetiology of the hydrocephalus.
- MRI. This may demonstrate the anatomical features as the CT scan, but offers particular clarity for lesions around the third ventricle and for aqueduct stenosis.
- Ultrasound scan. This may be used while the anterior fontanelle remains patent.

Management

The underlying cause of the hydrocephalus should be treated, where possible, but in the majority of cases a shunt is also necessary to divert CSF from a lateral ventricle into the peritoneal cavity. Ventriculoatrial shunts (diverting flow into the right atrium) are now less frequently used. Flow of CSF is controlled by valves in the tubing which prevent reflux of CSF and control the pressure at which CSF is released (Spitz–Holter or Pudenz types or their variants).

Drug therapy to reduce CSF production (acetazolamide, isosorbide dinitrate) usually has a short-lived effect and is of little value.

Complications of CSF shunts

Epilepsy occurs in 20–35% of patients with shunts. Shunt revision may be necessary as the child grows, either electively or because of recurrent symptoms. However, the common complications of shunts are:

1. Blockage. The shunt may block at either end, with obstruction by adhesions, omentum or the brain substance itself. There may be valve malfunction within the tubing. Symptoms of acute or chronic hydrocephalus occur, and urgent neurosurgical assessment is needed.

2. *Infection.* This tends to occur within weeks of shunt insertion, and usually results from perioperative shunt contamination. *Staphylococcus epidermidis* or *Staphylococcus aureus* is commonly responsible. Symptoms of malaise, vomiting and pyrexia occur, with an associated neutrophil leucocytosis and raised ESR or CRP. Diagnosis involves sampling of CSF from the shunt reservoir.

Treatment involves removal of the shunt and insertion of an external ventricular drain, and so such sampling should only be performed after discussion with the centre where the shunt was inserted. Intraventricular and intravenous antibiotics are also required.

Further reading

Guazzo EP. Recent advances in paediatric neurosurgery. *Archives of Disease in Childhood,* 1993; **69:** 335–8.

Hull J, Harkness W. Common complications of cerebrospinal fluid shunts. *Current Paediatrics,* 1992; **2(2):** 77–9.

Related topics of interest

Cerebral palsy (p. 56)
Congenital infection (p. 98)
Developmental delay and regression (p. 115)
Malignancy in childhood (p. 217)
Seizures (p. 288)

HYPERTENSION

Blood pressure increases with age and there is no precise definition of hypertension, but a child with a blood pressure greater than the 95th centile for age may be considered to have significant hypertension. Severe hypertension has been defined as a blood pressure greater than the 99th centile for age. Hypertension should not be diagnosed until an abnormal blood pressure has been confirmed on several occasions. Mild, asymptomatic hypertension, particularly in older children, is usually primary or essential. These children may be at increased risk of ischaemic heart disease and cerebrovascular disease in adult life. The majority of children with severe symptomatic hypertension have an underlying cause, most commonly renal parenchymal disease.

Problems

- Heart failure.
- Acute hypertensive encephalopathy.
- Increased risk of ischaemic heart disease and cerebrovascular disease in adult life.

Aetiology

1. *Primary (essential).*

2. *Secondary*

(a) Renal (95% of secondary hypertension)
- Chronic GMN.
- Reflux nephropathy.
- Haemolytic uraemic syndrome.

(b) Vascular
- Coarctation of the aorta.
- Renal artery stenosis.
- Renal vein thrombosis.

(c) Endocrine
- Cushing's syndrome.
- Primary hyperaldosteronism (Conn's).
- Congenital adrenal hyperplasia.
- Neuroblastoma.

(d) Other
- Intracranial tumours.
- Post head injury (cerebral oedema, intracranial haemorrhage).
- Drugs, e.g. steroids.
- Lead poisoning.
- Porphyria.

Clinical features

Hypertension is often asymptomatic and detected only on routine examination. Blood pressure measurement is an essential part of the examination of any child with renal, cardiac or endocrine disease. Three or more readings should be taken with a cuff whose bladder width is at least 75% of the upper arm length. A cuff that is too small will give a spuriously elevated blood pressure. In infants and small children the Doppler method is recommended.

Symptoms only occur as a result of the complications of hypertension – raised intracranial pressure (vomiting, headaches, visual disturbance, seizures) or left ventricular failure (dyspnoea, cough).

There may be a history of genitourinary symptoms suggesting a renal cause. The child should be examined for evidence of abdominal masses, endocrine disease, coarctation, renal bruits (renal artery stenosis), raised intracranial pressure and Turner's syndrome. Left ventricular hypertrophy and optic fundi changes develop in long-standing hypertension. Facial palsy is an uncommon presenting feature.

Investigation

- Urinalysis and urine culture.
- U&E and creatinine.
- Renal imaging (US, isotope scanning).
- Urinary catecholamines (vanillylmandelic acid, homovanillic acid).
- Other tests. Depending on history, examination and preliminary laboratory findings, e.g. plasma renin and aldosterone, cortisol, abdominal angiography, brain imaging.

Management

All patients with essential hypertension should be advised not to smoke, and to lose weight if obese. Girls should be advised against using the contraceptive pill. If the hypertension is secondary, it may be possible to treat the cause (e.g. repair of coarctation, relief of renal artery stenosis). If borderline, the hypertension should be monitored and drug treatment may not be necessary. If the blood pressure is undoubtedly high or if there are signs or symptoms the hypertension should be treated with hypotensive drugs (diuretics, beta-blockers, vasodilators, angiotensin-converting enzyme inhibitors).

Acute hypertensive encephalopathy

Acute hypertension is most commonly seen in glomerulonephritis, haemolytic uraemic syndrome or following a head injury. Encephalopathy may rapidly develop, with vomiting, headaches, neurological signs (e.g. visual symptoms, facial palsy), impaired consciousness and seizures. It is an emergency requiring controlled blood pressure reduction with i.v. sodium nitroprusside or labetalol.

Further reading

Deal JE, Barratt TM, Dillon MJ. Management of hypertensive emergencies. *Archives of Disease in Childhood*, 1992; **67:** 1089–92.
de Swiet M, Dillon MJ, Littler W *et al.* Measurement of blood pressure in children. Recommendations of a working party of the British Hypertension Society. *BMJ*, 1989; **299:** 497.
Report of the Second Task Force on Blood Pressure Control in Children. *Pediatrics*, 1987; **79:** 1–25.

Related topics of interest

Adrenal disorders (p. 14)
Chronic renal failure (p. 75)
Coma (p. 85)
Heart failure (p. 168)

HYPOGLYCAEMIA

Blood glucose concentration is normally maintained within narrow limits by homeostatic regulation of glucose production and utilization. Hypoglycaemia is a common problem in the newborn period, being particularly associated with prematurity and small size. It is less common in infancy and childhood, but may be the first indication of metabolic or endocrine dysfunction. The exact incidence is difficult to ascertain as there is no consensus over the definition of hypoglycaemia, with glucose concentrations ranging from < 1 mmol/l to < 4 mmol/l in various textbooks. Repeated hypoglycaemia is a recognized cause of neurological damage, but the level at which this occurs is unclear. Asymptomatic hypoglycaemia has been associated with abnormal sensory evoked responses at levels below 2.6 mmol/l.

Aetiology

Hypoglycaemia may result from either inadequate glucose production or excessive glucose consumption. In the neonatal period this may be transient or persistent.

1. Neonatal period and infancy

(a) Transient

- Decreased production. Poor oral intake, prematurity, small for dates.
- Increased utilization. Hyperinsulinism (infant of a diabetic mother, rhesus disease, maternal beta-blocker therapy, idiopathic).
- Both of the above. Fetal distress, hypothermia, sepsis, cyanotic CHD.

(b) Persistent

- Decreased production. Inborn errors of metabolism (e.g. glycogen storage disorders, fatty acid oxidation defects), hormone deficiencies (e.g. hypopituitarism, adrenal insufficiency).
- Increased utilization. Hyperinsulinism (nesidioblastosis, Beckwith–Wiedemann syndrome).

2. Childhood

- Decreased production. Liver disease (e.g. Reye's syndrome), toxins (e.g. alcohol), hormone deficiencies (e.g. growth hormone, cortisol), ketotic hypoglycaemia (accelerated starvation).
- Increased utilization. Hyperinsulinism (islet cell tumour, exogenous insulin excess in diabetes mellitus).

Clinical features

Symptoms occur as a result of the lack of glucose available for CNS metabolism and increased catecholamine secretion in response to hypoglycaemia. In the neonatal period features include irritability, jitteriness, apnoea, hypotonia and convulsions, although many infants are asymptomatic. In the older child features include lack of energy, tremor, headache, hunger, pallor, sweating, tachycardia, mental confusion, visual disturbance, behavioural abnormalities, seizures and coma. Hypoglycaemia in childhood usually occurs after a fast (e.g. overnight), but hyperinsulinism may precipitate hypoglycaemia after meals. Features such as hepatomegaly, cataracts and metabolic acidosis suggest an inborn error of metabolism. Short stature, micropenis or electrolyte disturbances suggest an endocrine disorder.

Hypoglycaemia in diabetes mellitus

Hypoglycaemia is a major hazard of insulin treatment in diabetes mellitus and becomes more common with the quest for better glucose control. It is caused by delayed or missed food, increased exercise or errors of insulin dosage. Children and parents should be aware of the causes, symptoms and treatment (oral sugar, i.m./s.c. glucagon, i.v. glucose). Recent reports have suggested that the use of human insulins may be associated with loss of awareness of hypoglycaemic symptoms, but this has yet to be proven. Hypoglycaemia seldom causes long-term problems, although prolonged and severe hypoglycaemia (usually associated with deliberate, massive insulin overdose) will result in cerebral oedema and may cause death.

Transient neonatal hypoglycaemia

Infants who are small for gestational age (SGA) have reduced glycogen stores which are rapidly used up after birth, and the blood glucose will fall over the first few hours of life. The risk of hypoglycaemia recedes once oral feeding is fully established, but there are a group of underweight children, particularly under the age of 3 years, who remain prone to ketotic hypoglycaemia owing to reduced gluconeogenic reserve. Premature infants are vulnerable to hypoglycaemia because of their limited glycogen reserves and immaturity of the gluconeogenic and glycogenolytic pathways. Birth asphyxia leads to depletion of glycogen reserves as a result of catecholamine and glucagon release. There may also be transient hyperinsulinism due to pancreatic ischaemia.

The transient hyperinsulinism of the infant of a diabetic mother usually resolves within the first week of life. The hyperinsulinism of Beckwith–Wiedemann syndrome may be transient or prolonged, occasionally persisting into adult life.

Persistent hyperinsulinism

Anatomical abnormalities of the pancreas (nesidioblastosis and islet cell adenomas) are the commonest causes of prolonged severe hypoglycaemia in infancy. Infants with nesidioblastosis are large for gestational age (LGA) and look similar to infants of diabetic mothers at birth. Hyperinsulinism is due to inappropriate pancreatic endocrine development, resulting in an excess of β-cells.

Investigation of hypoglycaemia

- Blood glucose.
- Blood insulin. If high or normal despite hypoglycaemia, this confirms hyperinsulinism.
- Urinary ketones. Hyperinsulinism causes non-ketotic hypoglycaemia. Hypoglycaemia with ketosis is a feature of hormone and enzyme deficiencies.
- Other investigations as indicated, e.g. for metabolic disorders or hormone deficiencies.

Management

1. Oral carbohydrate. Monitoring of blood glucose in high-risk neonates and early introduction of feeds may avoid the need for i.v. dextrose. Diabetics should have regular meals, take extra carbohydrate before vigorous exercise and be given oral glucose or dextrose at the first sign of hypoglycaemia.

2. Intravenous dextrose. Glucose utilization in normal infants is 5–7 mg/kg/min and in adults is 2 mg/kg/min. A child who needs more than these levels to maintain blood glucose has hypoglycaemia due to excessive utilization. Levels of 15–20 mg/kg/min are frequently needed to maintain normoglycaemia in the presence of hyperinsulinism.

3. Glucagon. Can be given intramuscularly or subcutaneously if i.v. access is difficult. It is useful in diabetics who are unable to take carbohydrate orally, but in other conditions it may lead to rebound hypoglycaemia by stimulating hyperinsulinism.

4. Diazoxide. Inhibits glucose-stimulated insulin release and is potentiated by simultaneous administration of a thiazide diuretic. It may be sufficient to control hypoglycaemia in Beckwith–Wiedemann syndrome.

5. Somatostatin long-acting analogues. May be useful in nesidioblastosis.

6. Surgery. For nesidioblastosis, insulinoma.

7. *Specific treatments,* e.g. hormone replacement, dietary management of inborn errors of metabolism.

Further reading

Koh THHG, Aynsley-Green A, Tarbit M, Eyre JA. Neural dysfunction during hypoglycaemia. *Archives of Disease in Childhood,* 1988; **63:** 1353–8.

LaFranchi S. Hypoglycemia in infancy and childhood. In: Mahoney CP, ed. *Pediatric and Adolescent Endocrinology. Pediatric Clinics of North America,* 1987; **34(4):** 961–82.

Lucas A, Morley R, Cole TJ. Adverse neurodevelopmental outcome of moderate neonatal hypoglycaemia. *BMJ,* 1988; **297:** 1304–8.

Related topics of interest

Diabetes mellitus (p. 118)

Small for gestational age, large for gestational age (p. 302)

IMMUNIZATION

Immunity may be achieved actively or passively. Active immunization is the basis of the routine childhood vaccination programme, using inactivated or attenuated live organisms or their products to stimulate antibody production (e.g. oral polio vaccine is a live attenuated vaccine). Protection by actively produced antibody or antitoxin will then last for months or years, with booster doses reinforcing immunity. Passive immunity is achieved by the injection of human immunoglobulin, providing immediate, but only temporary, protection.

The routine vaccination of children ensures the best possible protection from serious infectious disease, but relies on a positive, well-informed approach by all health-care workers to achieve optimum results.

The routine vaccination programme

The Department of Health and the Welsh Office have a standard immunization procedure, described below. Since October 1992, *Haemophilus influenzae* b (Hib) vaccine has been included in the schedule.

- Diphtheria, tetanus, pertussis, Hib, and oral polio vaccine is given at 2 months, 3 months and 4 months. This accelerated schedule was introduced in 1990 to follow the WHO guidelines, giving earlier protection against pertussis, and allowing less opportunity for intercurrent illnesses to delay immunization.
- Measles, mumps and rubella vaccine (MMR) is given at 12–18 months (any age over 12 months).
- Diphtheria, tetanus and oral polio booster is given immediately prior to school entry, age 4–5 years.
- Rubella (girls only) and BCG (certain health authorities) are given between 10 and 14 years. BCG may be given to Asian or other groups of high-risk neonates.
- Booster tetanus and oral polio is given at 15–18 years.

Prematurity is not a reason to postpone the immunization programme, and vaccines should be given at the appropriate chronological age, uncorrected for prematurity. Oral polio vaccine (live) should be deferred until just before discharge from a neonatal unit.

Contraindications to immunization

True contraindications to immunization are uncommon, and if doubts exist further advice should be sought from a consultant paediatrician, public health physician or district immunization coordinator.

General contraindications to immunization are:

- An acute illness (not a minor infection without pyrexia) at the time of vaccination.

- A severe local reaction to a previous dose of vaccine, i.e. involving erythema and swelling of most of the circumference of the upper arm, or most of the anterolateral surface of the thigh.
- A severe systemic response to a previous vaccine, i.e. fever of above 39.5°C within 48 hours of vaccination, anaphylaxis, prolonged screaming or convulsions within 72 hours of immunization.

Individual vaccines

1. Pertussis. Pertussis vaccine is a suspension of killed *Bordetella pertussis* organisms given as a triple vaccine combined with diphtheria and tetanus vaccines. Children with problem histories may need special consideration before proceeding to immunization with pertussis vaccine, e.g.

- Cerebral damage during the neonatal period.
- Evolving neurological conditions (but not stable conditions, e.g. Down's syndrome).

Many children with a personal or family history of convulsions or febrile convulsions may also be immunized against pertussis – they may be more at risk of convulsions, and these cases may need discussion with the paediatrician.

2. MMR. This combined vaccine consists of live attenuated measles, mumps and rubella vaccines, and was introduced into the routine schedule in 1988. It should be given irrespective of a previous history of the three illnesses, as diagnosis is often difficult. Specific contraindications are:

- Untreated malignant disease or altered immunity, including those children on immunosuppressive drugs or corticosteroids.
- True allergy to egg, neomycin or kanamycin.
- Immunization with another live vaccine (usually BCG) in the previous 3 weeks.

3. Polio. Live oral polio vaccine is routinely used in the UK, and contains live attenuated strains of polio vaccine. Inactivated polio vaccine is available for those children in whom a live vaccine is contraindicated. Symptoms of diarrhoea and vomiting are contraindications.

4. Haemophilus influenzae type b (Hib). Conjugate Hib vaccines have been developed, consisting of capsular polysaccharides linked to proteins, and Hib immunization is now included in the routine immunization schedule. General contraindications apply (see above), but there are no specific contraindications. Initial redness and swelling at the injection site may occur after the first immunization, but usually settles quickly.

5. *Influenza.* Immunization against influenza is strongly recommended for those at increased risk of complications, e.g. children with chronic respiratory, cardiac or renal disease.

6. *Other vaccines.* Vaccination against hepatitis A, hepatitis B, pneumococcus, typhoid, cholera, etc., may be required for certain children.

Further reading

Department of Health. *Immunisation against Infectious Disease.* London: HMSO, 1992.
Salisbury DM. The future for childhood immunisation. *Current Paediatrics,* 1993; **3(4):** 197–201.

Related topics of interest

Childhood exanthemata (p. 62)
Meningitis and encephalitis (p. 220)

IMMUNODEFICIENCY

Minor infections are very common during early childhood, representing a maturation of the immune response during a time of increasing contact with other children. However, further investigation is required if infections are unusually frequent, recurrent or severe or are associated with failure to thrive.

Primary defects of immunity may be considered in four main groups:

1. Defects of cell-mediated immunity. Fungal, viral, parasitic, mycobacterial and opportunistic infections are seen, e.g. *Cryptosporidium,* adenovirus, cytomegalovirus.

2. Antibody deficiency. Encapsulated bacterial pathogens, e.g. *Pneumococcus, Haemophilus,* lead to chronic sinopulmonary infection, meningitis, septicaemia and osteomyelitis.

3. Phagocyte dysfunction. Staphylococcal and fungal infections commonly affect the skin, mucous membranes, liver and bone.

4. Defects of complement function. C6, C7, C8 deficiency is associated with recurrent Neisserial infection. In deficiency of C1, C2 or C4, diffuse connective tissue disease is seen, with less frequent pyogenic infection.

Various immunodeficiency disorders are described below.

Severe combined immunodeficiency (SCID)

In SCID there is a functional failure of cellular and humoral immunity. There may be a family history of an unexplained infant death. Infants usually present before 6 months of age with diarrhoea, failure to thrive, pneumonia and mucocutaneous candidiasis. Pulmonary, gastrointestinal and skin invasion with multiple viruses, fungi, protozoa and bacteria is frequently seen. Graft-versus-host disease may be triggered by lymphocytes in blood transfusions, or by transplacental passage of maternal lymphocytes. Lymphoid system hypoplasia with absent tonsils is a feature.

Transient hypogammaglobulinaemia of infancy

This is a rare self-limiting defect due to delayed maturation of immunoglobulin production. There may be a family history of immunodeficiency. Levels of IgG are very low, and levels of IgA and IgM are slightly reduced until 6–18 months of age, then a spontaneous recovery is the rule.

X-linked agammaglobulinaemia

This is an inherited humoral immunodeficiency, and affected males have absent or low levels of immunoglobulins A, M, G and D and circulating B cells. Patients typically present with a wide spectrum of infections between 6 and 24 months, as placentally transferred maternal immunoglobulins provide protection during the first 3–6 months of life.

Common variable immunodeficiency

This includes a varied group of patients with hypogammaglobulinaemia, reduced ability to produce antibodies after antigenic stimulation and T-cell impairment. Recurrent sinopulmonary infections, pneumonia and bronchiectasis typically occur after the age of 5 years. Complications are arthropathy, gastrointestinal disease and autoimmune phenomena such as idiopathic thrombocytopenic purpura or Coombs-positive haemolytic anaemia.

Leucocyte adhesion deficiency

Delayed cord separation, umbilical sepsis and soft-tissue infection are typical features of this autosomal recessive disease. There is defective expression of three leucocyte adhesion molecules, and neutrophils are therefore defective in their properties of chemotaxis, aggregation and phagocytosis. Staphylococcal, fungal and Gram-negative organisms are frequent pathogens.

DiGeorge syndrome

This congenital immunodeficiency is characterized by hypocalcaemia (hypoparathyroidism), congenital heart disease (mostly right-sided aortic arch and tetralogy of Fallot) with abnormal facies and defects of T-cell function.

Wiskott–Aldrich syndrome

This is an X-linked condition in which eczema and thrombocytopenia with reduced platelet size is accompanied by lymphopenia and reduced T-cell function. Levels of IgA and IgE are usually raised, with normal IgG and low IgM levels.

Chronic granulomatous disease (CGD)

The neutrophils of patients with CGD are unable to produce antimicrobial oxidant factors to kill microorganisms, although bacterial engulfment occurs normally. Inheritance may be X-linked or autosomal recessive, and suppurative lymphadenopathy, osteomyelitis, hepatic abscesses and purulent skin lesions are frequent features.

HIV infection

See Related topics.

IgG subclass deficiency

Recurrent bacterial infections may occur if patients are deficient in one or more of the four subclasses of IgG, e.g. IgG2 subclass deficiency is associated with recurrent meningococcal and pneumococcal infection.

Selective IgA deficiency

This is the most common immunodeficiency (1 in 400 to 1 in 700). Many patients are asymptomatic, but there are associations with IgG subclass deficiencies, autoimmune phenomena, gastrointestinal disease, pulmonary infection and allergy.

Investigation of immunodeficiency

- Microbiology – parasitic, bacterial, fungal and viral pathogens (including HIV) must be identified.
- FBC, film, ESR.
- Direct Coombs' test.
- Immunoglobulin levels.
- Specific antibody measurement, e.g. to immunization antigens (measles, Hib), blood group antigens (isohaemagglutinins) and other antigens of pathogens recently encountered.
- Complement analysis. C3, C4, CH50.
- T-cell numbers (CD4/CD8 ratios).
- Bone marrow aspiration.

If confirmatory tests of immune pathways are needed to define the suspected defect, or if no clinical diagnosis is reached, specialized immunological laboratory help is needed, e.g. assessment of opsonization and chemotaxis, measurement of IgG subclass or adhesion protein expression (CD11/CD18).

Principles of management

Infections should be promptly treated, with narrow-spectrum agents wherever possible. Prophylactic agents generally increase the risks of resistant organisms, although cotrimoxazole is of benefit in preventing *Pneumocystis carinii* pneumonia in patients with HIV. Live vaccines should be avoided. If there is cellular immunodeficiency, transfusions of blood products should be irradiated to prevent graft-versus-host disease. Immunoglobulin replacement therapy should be given to patients with antibody deficiency and are best given intravenously every 2-4 weeks, depending on serum levels and clinical response. Bone marrow transplantion, ideally from an HLA-identical donor, is the treatment of choice for patients with SCID and DiGeorge syndrome. Human recombinant interferon has recently been used to increase phagocyte bactericidal activity. Prenatal diagnosis and detection of carriers is now available for selected immunodeficiency diseases.

Further reading

Roberton D. Basic investigation of children with recurrent infections. *Current Paediatrics,* 1993; **3(1):** 6–9.

Strobel S. Nonmalignant disorders of lymphocytes: primary and acquired immunodeficiency diseases. In: Lilleyman JS, Hann IM, eds. *Paediatric Haematology*. Edinburgh: Churchill Livingstone, 1992: 273–327

Related topics of interest

HIV infection (p. 171)
Immunization (p. 184)
Lymphadenopathy (p. 214)

INBORN ERRORS OF METABOLISM

An error may occur in any metabolic pathway as a result of a genetically inherited enzyme defect. A characteristic phenotype results, and many of the defects can now be identified by enzyme assays or by DNA technology. Most are recessively inherited and the majority are uncommon, but they may cause severe neurological impairment and are often fatal. An enzyme deficiency results in organ damage by:

- Accumulation of the enzyme substrate, e.g. hyperphenylalaninaemia in phenylketonuria (PKU), glycogen as a result of defects of the glycogen degradation pathway.
- Overproduction of other metabolites by metabolism of the substrate via alternative pathways, e.g. metabolic acidosis and hyperammonaemia in urea cycle disorders.
- Lack of production of the normal metabolites, e.g. hypoglycaemia in disorders of fatty acid oxidation.

Treatment by induction of enzyme activity or by biochemical correction with special diets may improve symptoms and allow normal development in some cases. Identification of the defect is important not only for treatment, but also for genetic counselling.

Presentation

Metabolic disorders can present at any age. They are more common in consanguineous marriages. There may be a family history of the condition or a history of a previous stillbirth or infant death. Onset may be insidious or the child may be completely well for months or years until a stressful event (e.g. prolonged fasting, infection) triggers a metabolic crisis which mimics Reye's syndrome (acute encephalopathy with hepatic dysfunction).

Typical presenting features include:

- Neonatal symptoms. Hypotonia, drowsiness, fits, vomiting, hypoglycaemia and metabolic acidosis. These infants are often initially thought to be septicaemic. Symptoms improve on i.v. fluids but worsen again on reintroduction of feeds.
- Sudden unexpected infant death.
- Failure to thrive.
- Dysmorphic features, cataracts.
- Peculiar smell.
- Episodic vomiting, lethargy, hypoglycaemia, metabolic acidosis.
- Muscle weakness or cramps, myoglobinuria.
- Cardiomyopathy.
- Hepatocellular damage/visceromegaly.
- Developmental delay or regression.

- Convulsions.
- Acute encephalopathy.

Amino acid disorders

Symtoms occur due to toxic accumulation of amino acids or their metabolites. Untreated phenylketonuria (absent phenylalanine hydroxylase activity) has early adverse effects on the developing CNS, and neonatal screening is routinely performed in the UK (Guthrie test). Tyrosinaemia (hepatic toxicity), non-ketotic hyperglycinaemia (early CNS toxicity) and homocystinuria (skeletal, CNS, eye and vascular effects).

Organic acid disorders

These are disorders of the metabolism of branched-chain amino acids (leucine, isoleucine and valine) which present with metabolic acidosis and characteristic urinary organic acids, some of which have distinctive odours. When the infant is encephalopathic, hyperammonaemia and profound ketonuria are usual.

Examples include methylmalonic acidaemia, propionic acidaemia and maple syrup urine disease.

Disorders of fatty acid metabolism

Fatty acids are metabolized by β-oxidation within mitochondria. Carnitine is required for transport of long-chain fatty acids into mitochondria. Defects result in hypoglycaemia, abnormal urinary organic acids, and hyperammonaemia. Examples include carnitine deficiency and medium-chain acyl-CoA dehydrogenase (MCAD) deficiency.

Urea cycle defects

Defects may occur at any step in the metabolism of waste nitrogen to urea. Encephalopathy is associated with respiratory alkalosis with or without metabolic acidosis, hyperammonaemia + + +, abnormal urine and plasma amino acids especially glutamine (raised urinary orotic acid in ornithine carbamoyl transferase (OTC) deficiency). Examples include OTC deficiency, citrullinaemia.

Disorders of pyruvate metabolism

Pyruvate is the end-point of anaerobic glycolysis and is the starting point of gluconeogenesis. Lactate arises from normal anaerobic glycolysis and under normal conditions is in equilibrium with pyruvate (L:P = 15–25:1). The lactate to pyruvate ratio in urine, plasma and CSF increases in hypoxaemia, and in inherited disorders of the respiratory chain. Other findings include hypoglycaemia, metabolic acidosis, abnormal plasma and urinary amino acids. An example is hereditary lactic acidosis.

Disorders of carbohydrate metabolism

Defects in the glycolytic pathway result in hypoglycaemia, with excessive proteolysis and fatty acid oxidation to compensate for the lack of glucose. This leads to the accumulation of metabolites such as pyruvate, lactate and ketone bodies (metabolic and lactic acidosis). Examples include fructose-1,6-diphosphatase deficiency and galactosaemia. Galactosaemia presents with cataracts and severe liver disease in the neonatal period and is due to galactose-1-phosphate uridyl transferase (Gal-1-put) deficiency.

Acute metabolic encephalopathy

Inborn errors may cause acute encephalopathy by disturbance of glucose metabolism, metabolic acidosis, hyperammonaemia or liver dysfunction.

Investigation

- Blood glucose.
- Acid–base status.
- Liver function tests + clotting studies.
- Plasma ammonia. Normally < 50 μmol/l but will rise in any unwell child. Encephalopathy may occur at a level > 100 μmol/l, but it is usually over 200 μmol/l. Other causes of hyperammonaemia include Reye's syndrome, liver failure, proteus urinary tract infection, drugs (e.g. asparaginase, sodium valproate).
- Urinary ketones.
- Store frozen plasma, urine ± CSF for further diagnostic tests, e.g. urinary amino and organic acids, plasma amino acids. Definitive diagnosis may require enzyme measurement in cultured skin fibroblasts or DNA analysis (e.g. MCAD deficiency).

Management

1. Correct hypoglycaemia with dextrose infusion, ensure high energy intake to restrict catabolism and exclude protein.

2. Treat shock and maintain fluid balance.

3. Sterilize gut with oral antibiotics and laxatives to reduce enterohepatic recirculation of nitrogen.

4. Treat raised intracranial pressure due to cerebral oedema, and control seizures.

5. Specific therapies to remove toxins, e.g. L-arginine enhances the urea cycle, and sodium benzoate or sodium phenylbutyrate improve ammonia excretion. Dialysis is sometimes used to remove toxins and maintain fluid balance.

Outcome

The outcome of acute metabolic encephalopathy is usually poor so it is important to recognize inborn errors of metabolism before this occurs.

General management of inborn errors

1. Exclusion diets. Reduce dietary intake of enzyme substrate and replace essential nutrients. Enhance excretion of metabolites, e.g. ammonia. Some conditions are vitamin responsive.

2. Emergency regimens. During intercurrent illness change diet to a high carbohydrate intake, e.g. soluble glucose polymer drinks, to prevent decompensation and encephalopathy. If the child is vomiting, refuses to drink or is becoming encephalopathic change to i.v. dextrose.

3. Enzyme replacement therapy, e.g. bone marrow and organ transplantation, have been used in some conditions (Hurler's syndrome)

4. Genetic counselling and antenatal diagnosis. There is a one in four recurrence risk if the mode of inheritance is autosomal recessive. Enzyme assays or DNA analysis may be possible on amniotic cells or chorionic villus samples.

5. Screening, e.g. phenylketonuria (PKU) on Guthrie test.

Further reading

Dixon MA, Leonard JV. Intercurrent illness in inborn errors of intermediary metabolism. *Archives of Disease in Childhood,* 1992; **67:** 1387–91.

Green A, Hall SM. Investigation of metabolic disorders resembling Reye's syndrome. *Archives of Disease in Childhood,* 1992; **67:** 1313–17.

Wraith. Diagnosis and management of inborn errors of metabolism. *Archives of Disease in Childhood,* 1989; **64:** 1410–15.

Related topics of interest

Antenatal diagnostics (p. 23)
Coma (p. 85)

JAUNDICE

Jaundice is clinically detectable when serum bilirubin exceeds 85 μmol/l and, in the older infant or child, may indicate serious underlying pathology. Jaundice may be unconjugated (conjugated fraction less than 10–15%, bilirubin absent from the urine) or conjugated (conjugated bilirubin exceeds 15%, and bilirubin present in the urine). Unconjugated hyperbilirubinaemia may be physiological in the newborn, but conjugated hyperbilirubinaemia is always pathological.

Causes of unconjugated hyperbilirubinaemia

- Haemolysis. Most common.
- Impaired bilirubin clearance. In the newborn or due to familial hyperbilirubinaemias.

Causes of conjugated hyperbilirubinaemia

- Hepatocellular disease, e.g. viral hepatitis.
- Extrahepatic biliary obstruction, e.g. biliary atresia.
- Intrahepatic biliary hypoplasia (e.g. Alagille's syndrome) or obstruction.
- Dubin–Johnson or Rotor syndromes. Rare.

Clinical assessment

The history should always elicit whether family members have had jaundice or liver disease, and about consanguinity. There may be a recent history of an anaesthetic, contact with infectious disease, blood transfusion or travel. Jaundice can be seen in the conjunctivae, mucous membranes and skin. Further examination may reveal pyrexia, hepatic tenderness or enlargement, splenomegaly or stigmata of chronic liver disease. Stool and urine colour should be inspected.

Investigation

- FBC, reticulocyte count, blood group and Coombs' test.
- Coagulation screen.
- Liver function tests, including total and conjugated bilirubin, albumin, globulin, transaminases, alkaline phosphatase and γ-glutamyl transferase.
- Blood culture; viral serology.
- Urine for presence of bile; viral and bacterial culture.
- Abdominal ultrasound. To assess liver size and texture, portal and hepatic venous flow, presence of biliary obstruction, ascites. Other imaging with CT or MRI scan, isotope scans and rarely ERCP may be appropriate.
- Autoantibodies, immunoglobulins, liver biopsy. Occasionally required.

Viral hepatitis

Hepatitis A

Hepatitis A is an RNA virus, spread by the faecal–oral route, particularly in situations of poor sanitation or low socioeconomic conditions. The incubation period is 15–40 days, and jaundice may be preceded by nausea, vomiting, pyrexia, diarrhoea and abdominal pain. Diagnosis is confirmed by the appearance of IgM antibodies to hepatitis A. Complications are a prolonged cholestatic phase, fulminant hepatic failure and rarely bone marrow aplasia. There is no specific treatment – antiemetics should be avoided because of impaired hepatic metabolism, and severe vomiting may be a warning sign of fulminant hepatic failure. Passive protection may be given by gammaglobulin.

Hepatitis B

This DNA virus may be spread perinatally from a mother carrying 'e' antigen for hepatitis B, or from a mother acutely infected. It may also be transmitted via infected blood products, intravenous drug abuse and occasionally from body secretions in institutionalized children, who may bite and scratch. The incubation period is 3–6 months. Infection may cause acute hepatitis, fulminant hepatic failure or asymptomatic development of the hepatitis B virus (HBV) carrier state. Other features include papular dermatitis, chronic active or persistent hepatitis, and an association with primary hepatocellular carcinoma. Diagnosis of hepatitis B infection depends on the appearance of hepatitis B surface antigen 3-6 weeks after infection. IgM anticore, HBcAb, is the first antibody to appear, then HBeAb signals the end of active viral replication, and finally, HBsAb indicates recovery and protection from reinfection. Vaccination with hepatitis B vaccine is advisable for high-risk individuals.

Many other viruses e.g. herpes, adenovirus, Epstein–Barr virus, cause hepatitis as part of a generalized infection.

Hepatitis C

This is an enveloped, single-stranded RNA virus which is transmitted parenterally. It accounts for 95% of post-transfusion hepatitis. Infection may be asymptomatic, or, after an incubation period of 7–8 weeks, it may cause an acute hepatitis, with chronicity in up to 70% of cases (chronic active hepatitis, chronic persistent hepatitis, cirrhosis).

Inherited conditions

Gilbert's syndrome

There is impaired bilirubin clearance owing to impaired activity of uridine diphosphate glucuronyl tranferase (UDPGT), and probably impaired hepatic uptake. Prevalence is 6%, and inheritance is autosomal dominant with variable expression. A mild hyperbilirubinaemia, <70 μmol/l, typically presents after puberty in the absence of other significant symptoms, or the raised bilirubin may be found coincidentally on biochemical testing. Fasting and intercurrent infections elevate the bilirubin. Evidence of haemolysis and significant liver disease is absent. A liver biopsy is not routinely indicated.

Crigler–Najjar type 1

This rare autosomal recessive condition causes complete absence of UDPGT and presents as severe unconjugated neonatal jaundice (bilirubin levels > 340 μmol/l), progressing to kernicterus. There is no response to phenobarbitone, and prolonged phototherapy is required to control bilirubin levels. Liver transplantation should be considered.

Crigler–Najjar type 2

This type is also rare. There is decreased or absent UDPGT, usually causing jaundice in infancy. Bilirubin levels are lower, usually < 340 μmol/l, and fall with phenobarbitone. Phototherapy is only occasionally required, and neurological impairment is uncommon. Inheritance is autosomal dominant with incomplete penetrance.

Further reading

Gregorio GV, Mieli-Vergani G, Mowat AP. Viral hepatitis. *Archives of Disease in Childhood,* 1994; **70:** 343–8.
Tanner S. *Paediatric Hepatology*. Edinburgh: Churchill Livingstone, 1989.

Related topics of interest

Liver disease (p. 208)
Neonatal jaundice (p. 231)

KAWASAKI DISEASE

Kawasaki disease (mucocutaneous lymph node syndrome) is a systemic vasculitis of early childhood, with a prevalence of 3.4 per 100 000 children under 5 years in the British Isles. Oriental races are more at risk than Caucasian (80 000 reported cases in Japan), and although epidemiological studies are highly suggestive of an infective aetiology no definite pathogen has yet been established. Sudden death from coronary complications occurs in at least 2% of cases, and although specific therapy with aspirin and intravenous immunoglobulin may reduce coronary artery abnormalities, treatment must be given during the initial 10 days of the illness to be of benefit. Prompt diagnosis is therefore essential.

Epidemiology

- Eighty per cent of patients are under 4 years old.
- Spring and summer outbreaks occur in the UK, winter and spring outbreaks in Japan.
- Epidemic waves of outbreaks in different regions over months, with peaks every 3 years.
- Male:female ratio 1.5:1.
- No genetic links, but racial predisposition (Orientals).

Pathogenesis

There is a vasculitis of small and medium-sized arteries throughout the body. Damaged vessels may develop aneurysms and thrombosis, and may heal with fibrosis and calcification, initiating premature atherosclerosis. The mechanism of endothelial injury is unclear but may be an abnormal reaction to a common infective agent.

Clinical features

Diagnosis is established by the presence of five out of six characteristic clinical criteria and the exclusion of conditions which may mimic Kawasaki disease. There is no single or specific diagnostic test. The criteria are:

(a) Fever of 5 or more days duration.
(b) Presence of four of the following five conditions:
- Bilateral conjunctival injection.
- Change(s) in the mucous membranes of the upper respiratory tract, e.g. injected pharynx, dry, cracked lips or strawberry tongue.
- Change(s) in the peripheral extremities, e.g. oedema, erythema or desquamation.
- Polymorphous rash.
- Cervical lymphadenopathy.

(c) Exclusion of staphylococcal and streptococcal infection, measles, drug reaction, juvenile rheumatoid arthritis, etc.

In the presence of coronary artery aneurysms, fever plus three of the four criteria in (b) is sufficient for diagnosis.

Fever is usually high, spiking and prolonged, lasting for 1–2 weeks before subsiding or persisting as a low-grade fever for a further 2–3 weeks. The characteristic desquamation of the fingers and toes is not seen until 10–20 days after the onset of the fever. The rash may be extremely variable, with urticarial, maculopapular or multiforme-type lesions of various distributions around the body. Perineal desquamation may also be seen.

Other features may be seen, such as irritability, arthritis, aseptic meningitis, hepatitis and hydropic gall bladder.

Cardiovascular complications occur in 20–30% of patients. Myocarditis, pericarditis, arrhythmias and cardiac failure may occur in the first 10 days of the illness. Arteritis of the coronary vessels may lead to aneurysm formation, with a peak frequency 4 weeks after the onset of symptoms. Sudden death may occur from massive myocardial infarction or haemopericardium after aneurysmal rupture between 2 and 12 weeks of disease onset.

Investigation

- FBC. Normochromic normocytic anaemia and polymorphonuclear leucocytosis are common in the first week. The platelet count is often normal in the first week, but rises to very high peaks of 800 000 up to 2 000 000 x 10^9 by the third week.
- Acute-phase proteins. Raised ESR, CRP.
- Microbiology. Exclude other possible causes, e.g. by throat swab, blood cultures and viral titres.
- ECG.
- Chest radiograph.
- Echocardiogram. Full assessment should be performed by an experienced paediatric cardiologist in order to detect cardiac aneurysms.

Treatment

Once the diagnosis is established, treatment should commence without delay.

High-dose intravenous immunoglobulin should be given at a dose of 2 g/kg over 12–18 hours. High fluid volumes may be needed, and careful monitoring is essential if there is already evidence of cardiac involvement. Doses of 400 mg/kg/day for 4–5 days appear to be less effective.

High-dose aspirin (30 mg/kg/day) should commence immediately, and be continued until the fever has resolved. A lower dose (3–5 mg/kg/day) should then continue.

If the initial cardiac echo was abnormal, repeated assessments should be made at weekly intervals, with the addition of dipyridamole or prostacyclin if there are coronary aneurysms. If initial cardiac echo was normal, repeat assessment may be made at 6 weeks, as late aneurysm formation may still occur. If coronary artery abnormalities do occur, long-term follow-up with continuing low-dose aspirin therapy is advised.

Further reading

Dhillon R, Newton L, Rudd PT. Management of Kawasaki disease in the British Isles. *Archives of Disease in Childhood,* 1993; **69:** 631–8.

Levin M, Tizard EJ, Dillon MJ. Kawasaki disease: recent advances. *Archives of Disease in Childhood,* 1991; **66:** 1369–72.

Nadel S, Levin M. Kawasaki disease. In: David TJ, ed. *Recent Advances in Paediatrics,* Vol. 11. Edinburgh: Churchill Livingstone, 1993: 103–6.

Related topics of interest

Childhood exanthemata (p. 62)
Pyrexia of unknown origin (p. 282)

LEUKAEMIA AND LYMPHOMA

Leukaemia is the commonest childhood malignancy with 400–450 new cases per year in the UK. This gives a prevalence of 3.5 per 100 000 children under the age of 15 years. Acute lymphoblastic leukaemia (ALL) accounts for 75% of cases; the peak age is 2–6 years, and it is slightly more common in boys (1.2M:1F). Non-Hodgkin's lymphoma is the third most common childhood malignancy, and it can be thought of as 'lumpy leukaemia'. The peak age is 7–10 years and it is three times more common in boys than in girls. Hodgkin's lymphoma occurs in older children and is four times more common in boys.

Leukaemia

Presentation

Presenting features include fatigue, pallor, bruising, frequent or severe infections, lymphadenopathy or bone pain. Hepatosplenomegaly may be present. Cranial nerve palsies (particularly seventh nerve) are uncommon presenting features but indicate CNS involvement. There may be testicular enlargement in boys owing to leukaemic infiltration

Investigation

- *FBC and film.* Typically shows anaemia, thrombocytopenia, neutropenia and blasts, but the blood count may be normal. A high white count (> 50 x 10^9/l) is a poor prognostic indicator.
- *Bone marrow aspirate and trephine.* Myeloperoxidase or Sudan black reaction distinguishes lymphoblasts from myeloblasts (negative or less than 3% positive blasts in ALL). Immunological markers determine the cell lineage of the blasts (T, B, null, common, pre-B, etc.).
- *Cytogenetics.* Numerical or structural chromosomal abnormalities in blood or bone marrow are of prognostic importance, e.g. hyperdiploidy has a good prognosis, while some translocations are associated with a worse prognosis.
- *Lumbar puncture.* Less than 5% of affected children have CNS disease at diagnosis. If there is any suggestion of CNS involvement or raised intracranial pressure, a CT scan should be performed before LP.
- *Culture.* Blood, urine, CSF, surface swabs.
- *Chest radiography.* Mediastinal widening is seen particularly in T-cell ALL.
- *U&Es, creatinine, calcium, phosphate and urate.* Baseline pretreatment. There may be a degree of renal failure at diagnosis, and the urate is often raised.

- *Liver function tests* (LFTs) are usually normal at diagnosis but enzymes may be elevated due to leukaemic infiltration. Many of the anti-leukaemic drugs can impair liver function.
- *Viral antibodies* (varicella, measles, CMV).

Adverse prognostic features

- High white blood cell count (WBC).
- Age < 2 years or >10 years (survival is only 20% if over 16 years).
- Boys have a slightly worse prognosis than girls.
- Massive mediastinal widening.
- CNS disease.
- T-cell ALL, acute or chronic myeloid leukaemia (AML, CML).
- Chromosomal abnormalities, e.g. translocations.
- Slow responders (not in remission after 4 weeks' induction).
- Relapse during treatment.

Treatment

Most children with ALL in the UK are treated according to national protocols as part of the UK Acute Lymphoblastic Leukaemia (UKALL) trial. The current protocol (UKALL XI) consists of a 4-week period of remission induction followed by consolidation with two or three blocks of intensification. Regular intrathecal methotrexate, with or without high-dose intravenous methotrexate, is given as prophylaxis against CNS relapse. Following consolidation, a further 18 months' maintenance treatment (mainly oral chemotherapy) is given to complete a total of 2 years' treatment. Cranial irradiation is now reserved for children with high-risk disease (high WBC, CNS disease). Bone marrow transplantation from a compatible sibling is considered in first remission in poor-prognosis patients. If there are no adverse prognostic features, 5-year survival is greater than 70%, but in high-risk patients it is only 40–50%.

Problems

1. Tumour lysis syndrome. Acute renal failure may occur when treatment starts, owing to acute lysis of blasts, with release of uric acid and phosphate (especially if high WBC or B-cell disease). Prehydration, alkalinization of the urine and allopurinol promote renal excretion and should be commenced at least 24 hours prior to treatment.

2. Immunosuppression. Children with leukaemia are relatively immunosuppressed for the duration of their treatment and for at least 6 months after. They are particularly at risk of infections (viral and fungal as well as bacterial) following intensification blocks when the child is neutropenic. Granulocyte colony-stimulating factors may be used to stimulate bone marrow recovery from intensive chemotherapy. Antibiotic prophylaxis against the opportunistic infection pneumocystis pneumonia is given routinely. Measles and chickenpox can be life-threatening and immunoglobulin should be given if exposed and non-immune. Immunizations should not be given during and for 6 months after treatment as antibody responses will be inadequate and live vaccines may be life-threatening.

3. Side effects of cytotoxic drugs. The introduction of antiemetics such as ondansetron has greatly improved the tolerability of treatment. Other side-effects include alopecia, mucosal ulceration (e.g. methotrexate), renal toxicity (e.g. methotrexate), cardiotoxicity (e.g. anthracyclines) and peripheral neuropathy (e.g. vincristine).

Non-Hodgkin's lymphoma (NHL)

Forty per cent of these lymphomas are abdominal (usually B-cell), and 30% are mediastinal (usually T-cell). Other sites include head and neck lymph nodes and the kidneys. There is an increased incidence in immunodeficiency syndromes, e.g. Wiskott–Aldrich, ataxia telangiectasia, AIDS, and post organ transplantation. NHL is treated using protocols similar to those for ALL. Staging and prognosis depends on the site and number of groups of lymph nodes involved. Five-year survival is 85–90% in localized disease, but only 30–40% in advanced disease.

Hodgkin's lymphoma (Hodgkin's disease, HD)

Epstein–Barr virus may have a role in the aetiology of HD. Histology of affected lymph nodes shows characteristic Reed–Sternberg cells. The disease itself causes relative immunosuppression and there is a high risk of second malignancies. Localized disease is treated with radiotherapy, and more widespread disease is treated with chemotherapy. Overall the 5-year survival is over 80%, but it tends to be a lifelong relapsing and remitting disorder.

Further reading

Plowman PN, Pinkerton CR (eds) *Paediatric Oncology: Clinical Practice and Controversies.* London: Chapman & Hall, 1992: Chapters 9–12.

Related topic of interest

Malignancy in childhood (p. 217)

LIMPING AND JOINT DISORDERS

Limping or refusal to weight bear usually results from lower limb pain which may arise from a soft tissue, muscle, bone or joint problem. Joint disorders may be localized to a single joint (monoarthritis) or involve several joints (polyarthritis). All joints should be examined for evidence of pain, swelling or limitation of movement, and it should be remembered that hip pain is frequently referred to the knee. Limping without pain is less common, e.g. undiagnosed congenital hip dislocation.

Aetiology of a limp

- Soft tissue. Trauma, ingrowing toenail, nail in shoe, inguinal lymphadenitis.
- Muscle. Trauma, myalgia secondary to infection (e.g. viral), muscular dystrophy, early poliomyelitis, cramp.
- Bone. Trauma (accidental, non-accidental), osteomyelitis, tumour (leukaemia, osteosarcoma, neuroblastoma), rickets.
- Joint. Trauma, septic arthritis, haemarthrosis, juvenile chronic arthritis (JCA), tuberculous arthritis, irritable hip, Perthes' disease, slipped femoral epiphysis, Osgood–Schlatter disease, Henoch–Schönlein purpura, rheumatic fever, reactive arthritis (secondary to viral infection), intra-articular fracture, gut arthropathy (ulcerative colitis, Crohn's disease), Lyme disease.

Clinical features

There may be a history of trauma, but children frequently sustain bumps and falls which are often insignificant. Localized bone swelling, tenderness or heat suggests either osteomyelitis or a fracture. Enlarged inguinal lymph nodes secondary to soft-tissue inflammation can cause a child to limp.

Ankle, knee and hip joints should be examined for swelling, pain, and limitation of movement. Pain and swelling of the joints are features of septic arthritis, haemarthrosis (haemophilia, trauma), Henoch–Schönlein purpura, tuberculous arthritis and rheumatic fever. JCA causes swelling of one or more joints, and this may be painful or relatively pain-free. Irritable hip and slipped femoral epiphyses cause pain and limitation of movement without swelling. Limping without swelling or marked pain is a feature of Perthes' disease and undiagnosed congenital dislocation of the hip. *Note:* hip pain is frequently referred to the knee.

Pain may also be referred to the legs from the abdomen (e.g. psoas spasm in appendicitis) or from the spine. Evidence of associated conditions may be present, e.g. rash and haematuria in Henoch–Schönlein purpura, or gastrointestinal symptoms in gut arthropathy. Examination of a child with a limp should also include examination of the shoes (protruding nail, excessively tight).

Septic arthritis

Septic arthritis occurs mainly in children under 5 years and commonly affects the large joints (hip, knee, ankle, shoulder, elbow). One or more joints may be affected, with multiple sites being particularly common in the neonatal period. The child is usually toxic, unwell and resents movement. The affected joint is held in the position of maximum relaxation and is swollen, hot, red and painful.

Investigation

- FBC. Raised white count.
- ESR. Raised.
- Blood cultures.
- Aspiration of the joint. Microscopy, Gram stain and culture. *Staphylococcus aureus* is the commonest organism, but others include *Haemophilus influenzae, E. coli, Streptococcus, Pneumococcus, Salmonella* and *Meningococcus.*
- Radiology. There may be widening of the joint space. Adjacent osteomyelitis is particularly common in neonates as the primary site of infection is usually bone.

Management

1. Antibiotics. A high-dose antistaphylococcal agent (flucloxacillin) is combined with gentamicin or cephalosporin to cover non-staphylococcal infections. The minimum duration of treatment is 3 weeks, with administration being intravenous for at least the first week.

2. Pain relief. Aspiration of the joint and immobilization with traction or splinting will relieve pain. Analgesics may also be necessary.

3. Physiotherapy is commenced in the recovery phase to prevent contractures and strengthen muscles.

4. Complications. The prognosis is usually excellent, but complications can occur, particularly in neonates. Metastatic foci of infection in other joints and bones may occur in the acute phase. Bony destruction due to associated osteomyelitis may lead to pathological fractures and early osteoarthritis. Shortening of the limb or overgrowth of bone due to hyperaemia may cause a subsequent limp.

Irritable hip (transient synovitis)

This is the commonest cause of hip pain and limping in childhood. The aetiology is unclear, but it may follow a mild upper respiratory tract infection. Peak age is 2–10 years and it is more common in boys. An otherwise well child presents with sudden onset of a limp and pain in one hip or knee. Movements of the hip joint are limited, but there is no swelling. Investigations including FBC, erythrocyte sedimentation rate (ESR) and radiology are usually normal. The condition resolves in 7–10 days, and treatment consists of simple analgesia, and bed rest, with traction in some cases. A small proportion (<5%) subsequently develop features of Perthes' disease.

Perthes' disease

Perthes' disease presents with a limp and limitation of hip movement, particularly abduction. Pain and muscle spasm may occur but it is frequently pain-free. The majority are unilateral, peak age is 5–10 years, and it is much more common in boys. Episodes of segmental avascular necrosis of the femoral head occur over 1–2 years, leading to irregularity and fragmentation. Healing occurs over a further 2–3 years. The aetiology is unknown. When more than half of the epiphysis is affected femoral head deformities are common and these children are usually treated with traction, bracing or osteotomy to maintain the femoral head in the acetabulum during the healing phase.

Slipped femoral epiphysis

This condition is more common in boys and is associated with obesity, tall stature and gonadal immaturity. Peak age is 10–15 years and 20% are bilateral. Presentation is with pain in the hip, thigh or knee with no evidence of swelling. Radiographs of the hip joint show slippage of the capital epiphysis downwards and backwards. This may compromise the blood supply leading to avascular necrosis, and manipulation of the epiphysis increases this risk. Pinning is possible if the amount of slip is less than one-third of the diameter, but if greater than this osteotomy is carried out to realign the femoral head once spontaneous fusion has occurred. The other hip must be watched carefully.

Further reading

Jenkins EA, Hall MA. The limping child. *Current Pediatrics*, 1992; **2:** 223–5.

Related topics of interest

Haemophilia and Christmas disease (p. 159)
Osteomyelitis (p. 252)
Polyarthritis (p. 263)
Purpura and bruising (p. 278)

LIVER DISEASE

Fulminant hepatic failure (FHF)

In FHF, progressive mental deterioration is associated with severe impairment of hepatic function and, by definition, signs of encephalopathy are first seen within 8 weeks of the onset of liver disease. The syndrome has a high mortality of 65–90% depending on the degree of encephalopathy.

Problems

- Coagulopathy.
- Hypoglycaemia.
- Cerebral oedema.
- Hypoxia.
- Renal failure.
- Metabolic/acid–base impairment.
- Sepsis.

Causes of FHF

1. *Infective.* Viral hepatitis, Epstein–Barr virus, cytomegalovirus.

2. *Drugs and toxins.* Paracetamol, halothane, sodium valproate, isoniazid, *Amanita phalloides* (death cap mushroom) poisoning.

3. *Metabolic.* Tyrosinaemia, galactosaemia, Wilson's disease.

4. *Ischaemic.* Budd–Chiari syndrome, post shock or cardiac surgery.

5. *Neonatal hepatitis* of any origin may cause FHF in infancy.

Clinical assessment

Nausea, anorexia and persistent vomiting are accompanied by behavioural changes and coagulopathy. There may be a history of an infective or toxic hepatitis. Tachypnoea with deep respiratory effort is often an early feature of encephalopathy. Restlessness, aggression and confusion progresses to stupor and coma. Asterixis (flapping tremor of the outstretched hands) is a late feature. A search should be made for signs of chronic liver disease (palmar erythema, spider naevi, etc.). The degree of coma should be documented and graded, i.e. grades 1–4.

Investigation

- FBC, cross-match.
- Coagulation screen.

- U&E, glucose, acid–base, calcium, phosphate, magnesium.
- Liver function tests and amylase.
- Liver ultrasound scan.
- Viral serology.

Management

Management is supportive, aiming to avoid or minimize complications while awaiting spontaneous recovery of liver function or liver transplantation. Specialized intensive care is indicated. Sedative drugs should be avoided and protein feeds should be stopped, giving neomycin and lactulose. Coma depth should be monitored, e.g. by the Glasgow coma scale. Major bleeding may require intravenous vitamin K and fresh-frozen plasma, with H_2-blockers, e.g. cimetidine, to reduce haemorrhage from gastric erosions. Maintaining cardiorespiratory stability, normoglycaemia and acid–base balance is important, and sepsis should be vigorously treated. Methods to control intracranial pressure include hyperventilation and mannitol administration. Uncontrolled coagulopathy may preclude intracranial pressure monitoring. Transplantation offers the best chance of survival, but certain supportive treatments have been used, e.g. exchange transfusion, haemoperfusion.

Portal hypertension

The normal pressure in the portal venous system is 5–10 mmHg, and portal hypertension occurs if this pressure rises above 10 mmHg as a result of prehepatic, intrahepatic or post-hepatic obstruction of blood flow.

1. Prehepatic causes. Portal vein occlusion (neonatal occlusion or umbilical sepsis, extrinsic compression), increased blood flow.

2. Hepatic causes. Cirrhosis of any cause, cystic fibrosis, congenital hepatic fibrosis.

3. Post-hepatic causes. Budd–Chiari syndrome (hepatic vein occlusion), constrictive pericarditis, congestive cardiac failure.

Problems

- Gastrointestinal haemorrhage.
- Hypersplenism.
- Ascites.
- Hepatic encephalopathy.
- Septicaemia.

Clinical assessment

Portal hypertension frequently presents with bleeding oesophageal varices, often with splenomegaly, hypersplenism and ascites. If the cause is intrahepatic, there may be cutaneous stigmata of liver disease with periumbilical venous shunts, ascites and hepatomegaly. Post-hepatic causes often present with prominent ascites and tender hepatomegaly. There may also be signs of the causative disease, e.g. Wilson's disease, constrictive pericarditis.

Investigation and management

Baseline investigations are as for fulminant hepatic failure. Chest radiography and cardiac ultrasound will exclude constrictive pericarditis. Endoscopy will confirm the presence of varices and associated gastritis, and allows sclerotherapy. Liver biopsy may be indicated to allow histological diagnosis. Rarely, coeliac angiography is indicated if portosystemic shunt surgery is contemplated.

For management of variceal haemorrhage, see Gastrointestinal haemorrhage (p. 142).

Liver transplantation

Patients with chronic liver disease and also selected patients with metabolic disease or fulminant hepatic failure should be considered early for transplantation. Current 1-year survival figures in UK centres are 70–80%. Absolute contraindications are metastatic malignancy, permanent neurological deficit and intractable extrahepatic sepsis. Complications are rejection, infection (bacterial, fungal, viral and *Pneumocystis carinii*), renal failure and vascular thromboses. Expert initial assessment and care of the child's general condition, especially the nutritional status, will minimize post-operative problems.

Further reading

Beath SV, Booth IW, Kelly DA. Nutritional support in liver disease. *Archives of Disease in Childhood,* 1993; **69:** 545–7.

Chiyende J, Mowat AP. Liver transplantation. *Archives of Disease in Childhood,* 1992; **67:** 1124–7.

Mowat AP. *Liver Disorders in Childhood,* 2nd edn. London: Butterworths, 1987.

Related topics of interest

Gastrointestinal haemorrhage (p. 142)
Jaundice (p. 195)
Neonatal jaundice (p. 231)

LOWER RESPIRATORY TRACT INFECTION

Acute infection of the lower respiratory tract affects one child in 10 each year, and is a common reason for hospital admission, particularly in the young infant. Although primary viral and bacterial infection is common, there may be intrinsic factors which predispose the child to severe infection, e.g. cystic fibrosis or immunodeficiency.

Problems

- Respiratory failure.
- Recurrent bronchoconstriction.
- Septicaemia.
- Pleural inflammation.
- Bronchiectasis.

Acute bronchiolitis

Respiratory syncytial virus (RSV) is usually the pathogen responsible for yearly epidemics of bronchiolitis, although parainfluenza and adenovirus may be involved. Clinical features include fever, coryza and dry cough, with apnoea in the young infant. Tachypnoea, intercostal recession, inspiratory crepitations and expiratory wheeze are typical examination findings. Recurrent cough and wheeze during the year after infection occur in up to 80% of infants.

Investigation

- Nasopharyngeal aspirate. Viral immunofluorescence usually confirms RSV early in the clinical course.
- Chest radiography reveals hyperinflation with patchy collapse and consolidation. Heart failure and pneumonia must be excluded.
- Oxygen saturation monitoring and blood gas estimation if clinically indicated.

Management

Supportive treatment involves correction of hypoxia and maintenance of fluid balance. Mechanical ventilation is needed in 1–5% of patients.

Specific antiviral treatment with ribavirin speeds resolution of respiratory symptoms, but probably does not reduce mortality. It may be used for severe bronchiolitis or if cardiorespiratory problems already exist.

Antibiotics are not indicated unless there are pre-existing cardiorespiratory risk factors or possible staphylococcal pneumonia. Systemic steroids, mist therapy and physiotherapy have no place in routine management. Ipratropium bromide relieves acute bronchoconstriction in some infants and the role of inhaled steroids is still unclear.

Pneumonia

Viral agents are frequently responsible, e.g. RSV, influenza, parainfluenza and adenovirus. Bacterial infection is more common in the older child – *Streptococcus pneumoniae, Mycoplasma pneumoniae, Staphylococcus aureus,* and *Haemophilus influenzae* infection may occur. Gram-negative organisms (*E. coli, Pseudomonas*) may cause pneumonia in the debilitated host. *Pneumocystis carinii* and measles may produce serious pneumonia if immune function is compromised, e.g. HIV infection, cytotoxic therapy.

Clinical features often include a prodrome of sore throat and coryza, followed by fever, cough, tachypnoea and abdominal or chest pain. Chest signs may indicate bronchopneumonia with widespread crepitations and decreased air entry or a lobar pneumonia with a dull percussion note and bronchial breathing.

Investigation

- Chest radiography confirms the extent and site of infection, the presence of complications, e.g. pleural effusion, lung abscess, and confirms resolution after recovery.
- FBC. Neutrophil leucocytosis is common.
- U&E. Severe pneumonia may cause hyponatraemia by inappropriate release of antidiuretic hormone (ADH).
- Blood cultures, sputum specimens, culture of pleural fluid and viral immunofluorescence. All may aid identification of the infective organism. Rapid bacterial antigen screen on blood or lung fluid may be used if antibiotics have already been used.
- Oxygen saturation monitoring and blood gas estimation if clinically indicated.

Management

Supportive treatment with oxygen, intravenous fluids and attention to nutritional status may be needed. Broad-spectrum antibiotics are appropriate until microbiology results define an organism. Occasionally, surgical drainage of an empyema or effusion is required.

Pertussis (whooping cough)

Epidemics of pertussis have previously occurred at 3- to 4-yearly intervals, but vaccine uptake has now increased again. *Bordetella pertussis* is the causative organism. Morbidity is significant (apnoea, cerebral hypoxia, bronchopneumonia and lung collapse), and deaths in young infants have occurred during recent epidemics. After a 2-week incubation period, a coryzal phase of 7–10 days is followed by the spasmodic phase, during which there are bursts of coughing without inspiratory pauses, followed by a sudden 'whoop' as air is drawn into the lungs. Vomiting, cyanosis, epistaxis and seizures may occur. Coughing may continue for many weeks – the 100 day cough.

Investigation

- FBC. Absolute lymphocytosis, usually above 20 000.
- Chest radiography. Bronchial wall thickening, secondary pneumonia, air trapping.
- Culture of *Bordetella pertussis* in respiratory secretions, or measure a rise in antibody titres.
- Monitor oxygen saturation and apnoeas.

Management

Expert nursing care during paroxysms is important, and supportive treatment with ventilatory support, fluids, nutrition, etc., may be needed. Erythromycin reduces the period of infectivity and should be given for 14 days. Oral salbutamol may reduce paroxysms. Infant vaccination should be promoted.

Tuberculosis

Mycobacterium tuberculosis is usually spread by aerosol, often from adult to child. The organism multiplies in the lung periphery, resulting in both acute and chronic inflammatory changes and calcification (Ghon focus). Spread to regional lymph nodes occurs, with hilar lymphadenopathy on the chest radiograph. Consolidation on chest radiographs may result from post-primary spread. Generalised spread may occur early (miliary tuberculosis) or late (to meninges, bone, kidneys). Attempts to culture the organism should be made via sputum or gastric washings. The Heaf test may be used for population screening, but the Mantoux test (delayed hypersensitivity reaction) is a more accurate test for previous infection. Treatment is with rifampicin and isoniazid for 6 months, supplemented with pyrazinamide during the first 2 months of therapy.

Further reading

Dinwiddie R. *Diagnosis and Management of Paediatric Respiratory Disease.* Edinburgh: Churchill Livingstone, 1990.

Related topics of interest

Asthma (p. 29)
Cystic fibrosis (p. 104)
Immunization (p. 184)

LYMPHADENOPATHY

Lymph node enlargement is a common finding during the clinical examination of children, and may be generalized or localized to a particular lymph node group. Commonly, lymphadenopathy is due to infection or local trauma. The main node groups are found in the anterior and posterior cervical triangle, occipital, submental, inguinal and epitrochlear regions and in the axillae. The underlying cause may be established on clinical grounds alone, but in some cases further investigation is required.

Causes of generalized lymphadenopathy

1. *Infection.* Generalized lymphadenopathy may be caused by bacterial, tuberculous or protozoan infections. Many viruses may also be responsible, including rubella, measles, cytomegalovirus, toxoplasmosis, Epstein–Barr virus and cat-scratch fever virus, or atypical mycobacteria.

2. *Malignancy.* Lymphoma, leukaemia, Langerhan's histiocytosis and metastases of solid tumours may produce generalized lymph node enlargement.

3. *Eczema.* Many children with generalized eczema have lymphadenopathy. Repeated scratching with secondary infection is probably responsible.

4. *Collagen disorders.* Lymphadenopathy may be a feature of juvenile rheumatoid arthritis and systemic lupus erythematosus.

5. *Drugs.* Certain drugs, e.g. carbamazepine and phenytoin, may occasionally be responsible for generalized lymphadenopathy.

6. *Storage disorders.* Lymphadenopathy (with hepatosplenomegaly) is a feature of Gaucher's disease and Niemann–Pick disease.

Causes of localized lymphadenopathy

1. *Localized infection.* Localized pyogenic infection (e.g. of throat, scalp, perineum, etc.) will cause lymph node enlargement in the group of nodes which drain the involved area.

2. *Metastases.* Metastatic spread of disease may be localized to nodes draining a particular region.

3. *Immunization.* Localized lymphadenopathy may occur after vaccination, e.g. BCG.

4. Lymphoma. Lymphoma may present with a localized enlargement of nodes.

Clinical features

A careful history should be taken, asking about systemic symptoms such as weight loss, anorexia, fever, sweats or malaise. Recent contact with infectious diseases should be noted and also any recent travel or drug ingestion. There may be a history of local trauma, sepsis or vaccination. An enquiry should be made about the appearance of rashes, found in eczema, various infectious diseases and collagen disorders.

On examination, the extent and nature of the lymphadenopathy should be defined. Splenomegaly may be present. The child may be pyrexial. Hard irregular nodes are often found in malignant disease, whereas acute infection often leads to tender nodes. The cause of localized lymphadenopathy may be apparent on careful examination, e.g. tonsillitis, perianal abscess. Other signs of specific diseases are sometimes found, e.g. bruising and purpura in acute leukaemia, signs of superior vena caval obstruction in lymphoma, or lichenified skin in chronic eczema.

Investigation

- FBC, film, ESR. The blood film may show atypical lymphocytes in infectious mononucleosis, or blast cells in haematological malignancy.
- Bacteriology specimens for culture and sensitivity, where relevant, e.g. from throat or wounds.
- Chest radiography. This may show hilar lymphadenopathy in sarcoidosis or tuberculosis, or a mediastinal mass with pleural or pericardial effusions in non-Hodgkin's lymphoma.
- A Mantoux test may be needed to exclude tuberculosis; also early morning urines and sputums for acid-fast bacilli. A differential Mantoux may be indicated to disguise atypical mycobacterial infections.
- Viral serology may assist in the investigation of rubella, toxoplasmosis, cytomegalovirus or Epstein–Barr virus.
- Bone marrow aspirate and trephine biopsy may prove necessary in certain cases to confirm haematological malignancy.
- Lymph node biopsy is needed if a satisfactory explanation for the lymphadenopathy is not apparent. This should be undertaken under a general anaesthetic, as pre-operative assessment may underestimate the involvement of underlying structures.

Related topics of interest

Immunodeficiency (p. 187)
Pyrexia of unknown origin (p. 282)

MALIGNANCY IN CHILDHOOD

Malignant disease is the second most common cause of death after accidents in children between 1 and 15 years, accounting for 16% of deaths. Nationally, approximately 1200 new cases of leukaemia and cancer are diagnosed each year. With a UK population of ~ 11 million children, this gives a 1 in 600 risk of developing cancer in the first 15 years of life. Overall, cancer is approximately one-third more common in boys. Despite being the second most common cause of death in children, a GP whose list includes 500 children could expect to see only two new cases in 35 years. However, by the year 2000, 1 in 1000 20-year-olds will be a survivor of childhood cancer, so the late effects of both the disease and its treatment are becoming increasingly important.

Problems

- Immunosuppression.
- Treatment tolerability.
- Family and social disruption.
- Late effects.
- Second tumours.

Relative incidences in the UK

1. Leukaemia (33%) is the commonest malignancy with 400–450 new cases per year (ALL accounts for 75% of cases).

2. Brain and spinal cord tumours (25%) are the second commonest childhood cancers with 300 new cases per year.

3. Lymphomas (11%). There are approximately 140 new cases per year, with non-Hodgkin's lymphoma being more common than Hodgkin's lymphoma.

4. Other solid tumours. Neuroblastoma (70 cases/year), Wilms' tumour (70 cases/year), retinoblastoma (35 cases/year), and hepatoblastoma (10 cases/year) together account for 15–20% of childhood cancers. There are approximately 60 new bone tumours (osteosarcoma, Ewing's sarcoma) and 50 new soft-tissue sarcomas (rhabdomyosarcoma) per year. Other tumours (e.g. germ cell, thyroid) are very rare.

Aetiology

The causes of childhood cancer are, for the most part, unknown but probably result from the interaction of a number of factors. The UK Co-ordinating Committee on Cancer Research (UKCCR) in collaboration with the UK Children's Cancer Study Group (UKCCSG) set up a case–control study in 1992 to investigate the aetiology of leukaemia. The study will try to establish the role of exposure of the parental germ cells, the fetus and the child to ionizing radiation and certain chemicals. Other possible aetiological factors include exposure to extremely low-frequency electromagnetic fields (e.g. proximity to high-power cables), or an abnormal response to a common infection.

In certain conditions there is a genetic predisposition to malignancy, e.g. chromosomal abnormalities (Down's syndrome), single gene defects (over 25% of children with retinoblastoma have a family history) and DNA repair syndromes (ataxia telangiectasia).

p53 is a tumour-suppressor gene located on chromosome 17. Raised levels and mutations have been found in many tumour types. Li–Fraumeni syndrome is described in families with germline mutations of *p53*, who have a much increased incidence of cancer, particularly breast cancer and soft-tissue sarcomas.

Treatment

Most children with cancer in the UK are now treated within clinical trials coordinated by the Medical Research Council (MRC), UKCCSG or European study groups. Results from these trials are continually evaluated and changes in protocols made to try to maximize survival whilst minimizing late effects. Without these national and international studies the numbers treated in each centre would be too small to detect statistically significant improvements in outcome rapidly. Improved survival has been achieved by the use of more intensive chemotherapy regimens which have been made possible by better supportive care (e.g. early introduction of antibiotics for febrile neutropenia, granulocyte colony-stimulating drugs to improve bone marrow recovery). The introduction of 5-HT_3 inhibitors (e.g. ondansetron) has reduced the severity of nausea and vomiting, and improved treatment tolerability.

Survival

Overall, more than 50% of children with leukaemia and cancer now survive beyond 5 years. For some malignancies (e.g. ALL, NHL, Wilms') the survival rates are now better than 70%, and in these it is important to minimize the late effects of the treatment. In contrast, survival rates for malignancies such as brain tumours and advanced neuroblastoma remain poor despite intensive chemotherapy and radiotherapy.

Late effects

The improved survival from childhood cancer has led to increasing concern about the long-term effects of the disease and its treatment on the survivors. These late effects include:

- Impaired growth and development (particularly following cranial irradiation).
- Increased risk of second tumours.
- Drug effects, e.g. cardiotoxicity secondary to anthracyclines.
- Infertility due to gonadal irradiation or chemotherapy.
- Neuropsychological sequelae which may be subtle defects (e.g. difficulties with mental arithmetic following cranial irradiation) or more pronounced, particularly when treated at a young age.

Treatment strategies must now be evaluated, not only in terms of improved survival, but also in terms of the health status of survivors.

Further reading

Chambers EJ. Radiotherapy in paediatric practice. *Archives of Disease in Childhood,* 1991; **66:** 1090–2.

Li FP, Fraumeni JF. Prospective study of a family cancer syndrome. *Journal of the American Medical Association,* 1982: **247:** 2692.

Morris-Jones PH, and Craft AW. Childhood cancer: cure at what cost? *Archives of Disease in Childhood,* 1990; **65:** 638–40.

Related topics of interest

Leukaemia and lymphoma (p. 201)
Solid tumours (p. 306)

MENINGITIS AND ENCEPHALITIS

Eighty per cent of all bacterial meningitis occurs in childhood, with a mortality of 10% (over 100 deaths per year in the UK). It is the commonest cause of acquired sensorineural deafness, and can lead to other neurological sequelae, including epilepsy. The organisms responsible vary with age, and the choice of antibiotics therefore depends on the age of the child as well as the local pattern of drug resistance. Meningitis may also follow infection with a wide variety of viruses, and the prognosis is usually good with no specific treatment. Encephalitis is usually viral or post-infectious and is characterized by fever, disturbed consciousness, convulsions and focal neurological signs. Tuberculous meningitis should be considered in every case of aseptic meningitis or encephalitis. The onset is usually insidious and these children often develop focal neurological signs before a diagnosis is made.

Problems

- Seizures.
- Inappropriate ADH secretion.
- Cerebral oedema.
- Subdural effusions.
- Septicaemic shock.
- Neurological sequelae.
- Prophylaxis for close contacts.

Meningitis

Aetiology

1. *Bacterial*

(a) Neonatal period
- *E. coli* and other Gram-negative organisms.
- Group B streptococcus.
- *Listeria monocytogenes.*

(b) Over 3 months
- *Neisseria meningitidis.*
- *Haemophilus influenzae* (usually under 6 years).
- *Streptococcus pneumoniae.*

(c) 1 to 3 months
- Any of the above.

2. *Viral.* Most commonly coxsackie, echovirus, mumps.

3. *Tuberculosis.*

4. *Fungal.* Rare unless immunocompromised.

Clinical features

The characteristic features are pyrexia, headache, neck stiffness and photophobia. In a neonate or infant the symptoms are less specific with fever, poor feeding, irritability, vomiting, a full fontanelle, apnoea or drowsiness. A convulsion may be the presenting feature, so a lumbar puncture (LP) should always be considered in a child under 1 year who has a convulsion with fever. Neonatal meningitis is uncommon but has a high mortality and morbidity; most have an associated bacteraemia and many are shocked at diagnosis. Meningococcal meningitis may be associated with septicaemia (characteristic rash ± shock). Haemophilus meningitis is less common since the introduction of routine immunization. Symptoms of viral meningitis are usually less severe than bacterial meningitis and there may have been a preceding upper respiratory or gastrointestinal illness, but it is not always clear-cut.

Investigation

- FBC, clotting studies.
- Blood cultures.
- Serum viral titres – acute and convalescent.
- Blood glucose.
- CSF protein, glucose, bacteriology and virology. In bacterial meningitis there is a raised WBC (mainly polymorphs), raised protein and low glucose (< 50% of blood glucose). Organisms may be seen on Gram stain. In viral meningitis there is a raised WBC (mainly lymphocytes), protein is normal or raised and CSF glucose is normal. In tuberculous meningitis there is a raised WBC (mainly lymphocytes), raised protein and low glucose.
- U&E.
- Swabs (throat, nose, purpuric skin) for bacterial and viral culture.
- Urine culture.

Management

1. Lumbar puncture. If meningitis is suspected, an LP should be performed immediately unless the child's cardiovascular system is unstable or there is evidence of raised intracranial pressure (e.g. deep coma, protracted seizures, focal neurological signs, unequal pupils or abnormal pupillary reflexes plus gradual onset of symptoms/signs). If the LP is contraindicated, antibiotic treatment should not be delayed. Antibiotic guidelines should be reviewed regularly to take account of the continually changing pattern of antibiotic resistance. Current recommendations suggest the initial use of ampicillin and cefotaxime in infants between 1 and 3 months, and cefotaxime alone for older infants and children. Cefotaxime, gentamicin and ampicillin are recommended in the neonatal period. Treatment should be reviewed with culture and sensitivity results after 24–48 hours. In the presence of a meningococcal rash penicillin should be given immediately. If the CSF and clinical picture suggests viral meningitis it is usual to treat with 48 hours' antibiotics, awaiting negative CSF culture.

2. *General measures.* Cardiopulmonary support may be necessary. Restrict fluid to two-thirds of normal requirements and record strict fluid balance. Inappropriate ADH secretion is common. Monitor neurological status – any deterioration suggests worsening cerebral oedema or development of a subdural collection. Detect and treat seizures.

3. *Steroids.* Much of the cerebral damage in bacterial meningitis is due to activation of the host inflammatory responses rather than the direct effect of the invading organism. The role of steroids in reducing this inflammatory response is controversial, but the early use of dexamethasone in haemophilus meningitis may be of benefit, particularly in reducing the incidence of deafness. The role of steroids in meningococcal and pneumococcal meningitis is still unclear, and their routine use is not recommended.

4. *Prophylaxis.* The disease must be notified to P.H.L. Close contacts (household or kissing contacts) of a child with meningococcal disease should be treated with rifampicin. The patient should also be given rifampicin to clear carriage as soon as oral intake is tolerated. Children under 5 years who have been in close contact with a child with haemophilus meningitis and who are not immunized should be given rifampicin. Rifampicin causes orange–red discoloration of urine and tears (stains contact lenses). It interferes with other drugs (including the oral contraceptive pill) and is contraindicated in pregnancy.

Encephalitis

Aetiology

1. *Infectious*
- Herpes simplex.
- Enterovirus (coxsackie, echovirus).
- Mumps.

2. *Post-infectious*
- Measles.
- Varicella zoster.
- Live vaccines (e.g. measles – very rare).
- Subacute sclerosing panencephalitis (SSPE).

Clinical features

Herpes simplex is the most common cause of severe encephalitis. Mortality is high and neurological sequelae in the survivors are common. The CSF shows a raised white count, and the EEG and CT scan show characteristic changes in the temporal lobes. An acute self-limiting encephalitis may follow chickenpox, resulting in cerebellar ataxia. Encephalitis may also follow measles, with irritability and seizures a few days after the rash. There may be complete recovery or it can progress to severe neurological impairment or death. SSPE is a late complication of measles owing to persistence of the virus in the brain. Degeneration of the brain occurs several years after the initial infection.

Further reading

Mellor DH. The place of computed tomography and lumbar puncture in suspected bacterial meningitis. *Archives of Disease in Childhood,* 1992; **67:** 1417–9.

Report of the Meningitis Working Party of the British Paediatric Immunology and Infectious Diseases Group. Should we use dexamethasone in meningitis? *Archives of Disease in Childhood,* 1992; **67:** 1398–401.

Gandy G, Rennie J. Antibiotic treatment of suspected neonatal meningitis. *Archives of Disease in Childhood,* 1990; **65:** 1–2.

Related topics of interest

Childhood exanthemata (p. 62)
Coma (p. 85)
Immunization (p. 184)
Neonatal convulsions (p. 227)
Shock (p. 291)

MUSCLE AND NEUROMUSCULAR DISORDERS

Disorders of muscle (myopathies) present with floppiness, delayed motor milestones, abnormal gait, clumsiness or progressive muscle weakness. The majority of myopathies are inherited, and all are uncommon. Dystrophies are characterized by degeneration of muscle fibres. Myotonia is the failure of muscle relaxation after contraction. Myotonia may occur as an abnormal response to muscle relaxant agents (e.g. scoline) during anaesthesia resulting in total opisthotonus. There is an increased risk of this reaction in children with muscle disease. Disorders of the neuromuscular junction include myasthenia gravis and botulism.

Myopathies

- Developmental, e.g. nemaline myopathy, central core disease.
- Degenerative, e.g. muscular dystrophy.
- Metabolic, e.g. glycogen storage disease (Pompe's), carnitine deficiency, periodic paralysis, mitochondrial abnormalities.
- Endocrine, e.g. hypothyroidism.
- Myotonic, e.g. dystrophia myotonica, abnormal response to scoline.
- Inflammatory, e.g. polymyositis, dermatomyositis.

Duchenne muscular dystrophy

Problems

- Progressive muscle wasting and weakness.
- Repeated respiratory infections.
- Psychological problems.
- Family disruption and modifications to the home.
- Genetic counselling. There is a 50% risk of recurrence in males.
- Future therapies. Gene replacement, myoblast transfer.

Clinical features

Duchenne muscular dystrophy is the most common muscular dystrophy with a prevalence of 1 in 3000 male births. It is inherited as a sex-linked recessive condition, although new mutations are responsible for up to one-third of cases. Female carriers are unaffected. Presentation is usually in the first 5 years of life with a waddling gait due to weakness of the pelvic girdle. Other early features include frequent falls, difficulty climbing stairs, climbing up the legs to get up from the floor (Gower's sign), pseudohypertrophy of the calf muscles and lordosis. The intellect is usually normal but there is an increased incidence of mild learning difficulties. Weakness and wasting is progressive with the majority unable to walk by 10 years. Cardiac muscle is also affected. Late features include scoliosis, contractures and aspiration pneumonia. Death usually occurs between 15 and 25 years. Becker muscular dystrophy is less common and progresses more slowly, with death usually in middle age.

Investigation

- Creatine phosphokinase (CPK). Very high (normal up to 150), falls as disease progresses.
- Neurophysiology. Characteristic myopathic changes.
- Muscle biopsy. Degenerative changes, dystrophin immunofluorescence.
- Cytogenetics. A very large gene involved in Duchenne and Becker muscular dystrophy has been located on the short arm of the X chromosome (Xp21). It codes for the protein dystrophin, which forms part of the cytoskeleton of muscle cells. Mutations can be identified in about 60% of cases. The level of dystrophin seems to correlate with clinical severity, and therapeutic measures to correct dystrophin abnormalities are the subject of current research.
- Antenatal diagnosis. Identification of a known mutation, or dystrophin analysis.

Myotonic syndromes

Dystrophia myotonica is the most common of the myotonic syndromes. Inheritance is autosomal dominant and the gene has recently been identified on chromosome 19. It was thought to be a disease of adult life, but onset in childhood does occur. Muscle weakness produces a myopathic facies, ptosis and a sagging jaw. Myotonia causes difficulty in relaxing the grasp. It is associated with mild learning difficulties. Congenital dystrophia myotonica occurs in infants of an affected mother and many die of respiratory difficulties in the neonatal period as a result of severe hypotonia. Those who survive often have severe learning difficulties. Myotonia is detected by shaking hands with the mother.

Dermatomyositis

This is an inflammatory disorder characterized by generalized proximal muscle weakness and pain (polymyositis) associated with a violaceous skin rash (eyelids, cheeks, knees and bony prominences). There is an underlying vasculitis of small blood vessels of unknown aetiology. Treatment is with steroids.

Myasthenia gravis

This is an autoimmune disorder of the neuromuscular junction which can occur at any age. Antibodies are produced against the acetylcholine receptors. It is associated with other autoimmune diseases and is more common in girls. Clinical features include ptosis, squint, difficulty in chewing and swallowing, and dysarthria. Generalized weakness and hypotonia is an uncommon presentation. Weakness is exacerbated by repetitive or sustained contraction. Diagnosis is confirmed by reversal of weakness with edrophonium and treatment is with long-acting anticholinesterase inhibitors. Acute exacerbation of muscle weakness can be precipitated by exertion, infections or some drugs (myasthenic crisis) or by excessive anticholinesterase treatment (cholinergic crisis). Neonatal myasthenia gravis occurs in some infants of affected mothers owing to placental transfer of antiacetylcholine receptor antibodies, and resolves in 4–6 weeks. Congenital myasthenia gravis is an autosomal recessive condition which presents in the newborn period and is relatively resistant to treatment.

Further reading

Bushby KMD. Recent advances in understanding muscular dystrophy. *Archives of Disease in Childhood,* 1992; **67:** 1310–2.

Related topic of interest

Floppy infant (p. 136)

NEONATAL CONVULSIONS

Many disorders may cause fits in the neonatal period, the commonest being perinatal asphyxia, metabolic derangements and meningitis. Asphyxia due to disturbances in fetal circulation and lack of oxygen during labour may lead to hypoxic–ischaemic encephalopathy in the first few days of life, and this is a major cause of death and neurological disability.

Aetiology

- Hypoxia. Perinatal asphyxia.
- Metabolic disorders. Hypoglycaemia, hypocalcaemia, hyponatraemia, hypomagnesaemia.
- Infection. Meningitis, congenital viral infection, encephalitis
- Intracranial haemorrhage. Intraventricular, subdural, intracerebral.
- Drug withdrawal. Maternal drug abuse.
- Cerebral malformation.
- Inborn error of metabolism.

Clinical features

Convulsions may be manifested by extension and stiffening of the body with upward deviation of the eyes. Other more subtle features include apnoea and bradycardia, sucking and chewing movements of the mouth or cycling movements of the limbs.

Investigation

- Blood sugar. BM Stix immediately.
- U&E.
- Serum calcium and magnesium.
- Acid–base status.
- Full septic screen including lumbar puncture.
- FBC.
- Other tests as indicated, e.g. metabolic screen, cerebral US, clotting screen, EEG.

Management

1. Correct underlying metabolic disorder, e.g. hypoglycaemia (i.v. dextrose infusion), hyponatraemia (fluid restriction). These may be the primary cause of the convulsion or may be secondary to other conditions such as birth asphyxia, meningitis or an inborn error of metabolism.

2. Anticonvulsants to control fits and prevent further convulsions. Phenobarbitone is the first-line drug. Other useful drugs include paraldehyde, phenytoin and clonazepam. If convulsions continue despite these, consider thiopentone.

3. Fluid balance. Cerebral insults (e.g. birth asphyxia, prolonged convulsions, meningitis) result in a degree of cerebral oedema. Fluids should be restricted by 20–40% in the first 24–48 hours to reduce the risk of fluid overload, which exacerbates cerebral oedema and may lead to hyponatraemia. Oliguria may result from ADH secretion or renal ischaemia. The use of diuretics (e.g. mannitol) and the place of intracranial pressure monitoring are controversial.

4. Respiratory support. Ventilation may be needed to maintain good oxygenation. Hyperventilation to maintain a low–normal $P\text{CO}_2$ may help reduce cerebral oedema.

5. Broad-spectrum i.v. antibiotics should be commenced once a full septic screen has been performed. The common causes of neonatal meningitis are group B streptococcus, *E. coli* and *Listeria.*

Birth asphyxia

Problems

- Hypoxic–ischaemic encephalopathy and neurological sequelae.
- Meconium aspiration.
- Cardiovascular complications – myocardial ischaemia, persistent fetal circulation.
- Renal ischaemia.
- Necrotizing enterocolitis.
- Inappropriate ADH.
- Disseminated intravascular coagulopathy (DIC).

Diagnosis and outcome

There is no accepted definition of birth asphyxia so the prevalence is difficult to ascertain, but in the UK it is in the region of 5 per 1000 full-term live births, with 1 in 1000 dying or being left severely disabled. Prompt cardiorespiratory resuscitation at birth and early supportive care (glucose, oxygen, management of cerebral oedema, treatment of seizures) are important in limiting the cerebral injury.

Hypoxic stress is an invariable consequence of labour, but the level of stress which an individual fetus can withstand is unclear. Meconium staining of the amniotic fluid and an abnormal cardiotocograph are currently used as indications of fetal distress during labour but are poorly predictive of long-term neurological problems. Fetal scalp blood sampling can be used to assess the degree of acidosis as a measure of fetal distress, and a pH < 7.2 is often taken as an indication for assisted delivery (forceps, Caesarean). The mean umbilical venous pH in uncompromised fetuses is reported to be 7.32, but there is poor correlation between cord blood acidosis and subsequent neurological impairment. The Apgar score is routinely recorded at 1 and 5 minutes after birth and assesses five variables: heart rate, respiratory effort, muscle tone, reflex irritability and colour. Severe depression of the Apgar score (0–3) is directly related to the risk of death, but it is otherwise poorly predictive of neurological outcome and there is a poor correlation with cord blood acidosis. Delay in establishing spontaneous respiration may indicate asphyxia but is influenced by gestational age and maternal drugs.

Three grades of hypoxic–ischaemic encephalopathy (HIE) following asphyxia are described: mild, moderate and severe. The duration and severity of neurological signs (level of consciousness, tone and posture, sucking reflex, seizures, autonomic function) together with EEG findings are the best predictors of neurological outcome. Mild encephalopathy carries a good prognosis, but moderate and severe encephalopathy result in a high risk of impairment. The severity of HIE, however, can only be diagnosed retrospectively after symptoms have developed. New therapeutic interventions such as allopurinol (a free radical scavenger) and inhibitors of the excitatory neurotransmitter glutamate may prove to be useful but need to be given early, so there remains a need for an early marker of asphyxia. Newer techniques such as magnetic resonance spectroscopy to evaluate the ATP energy state of the brain and Doppler measurement of cerebral blood flow may be useful in assessing the extent of asphyxia.

Ultrasound and CT scanning are used to identify treatable complications of asphyxia (e.g. subdural collection) and to predict outcome. Poor prognostic findings include injury to the basal ganglia and subcortical cysts due to infarction in the vascular watershed area, but these do not develop until after the first week of life.

Further reading

Levene MI. Management of the asphyxiated full term infant. *Archives of Disease in Childhood,* 1993; **68:** 612–6.

Marlow N. Do we need an Apgar score? *Archives of Disease in Childhood,* 1992; **67:** 765–7.

Related topics of interest

Cerebral palsy (p. 56)
Coma (p. 89)
Meningitis and encephalitis (p. 220)

NEONATAL JAUNDICE

Jaundice occurs commonly in the newborn period and, although it may be physiological and benign, it may also signify serious underlying disease.

Physiological jaundice

This occurs in healthy newborn infants. Jaundice is noted after the age of 24 hours, reaches a peak on the third or fourth day of life (approximately 120 μmol/l), and fades by the seventh to tenth day, with a later peak and trough in the preterm infant. Aetiological factors include the shortened red cell survival in neonates, reduced bilirubin conjugation and increased enteric reabsorption of bilirubin.

Breast milk jaundice

Jaundice is more common in breast-fed babies. This is thought to be due to the presence of β-glucuronidase in the milk, which deconjugates bilirubin in the bowel, causing increased enterohepatic circulation.

Pathological jaundice

Jaundice should always be investigated if there is deviation from the pattern of physiological jaundice, i.e. visible jaundice within 24 hours of birth, baby unwell, high or fast-rising levels of bilirubin (level above 220 μmol/l in a term baby, or a rate of rise above 85 μmol/l/day). Persisting jaundice beyond 7 days in a term baby or 14 days in the preterm baby deserves investigation. Unconjugated hyperbilirubinaemia is often physiological, but conjugated hyberbilirubinaemia (pale stools and dark urine) is always pathological.

Aetiology

1. Unconjugated hyperbilirubinaemia

- Haemolysis, e.g. rhesus disease or ABO incompatibility, red cell defects, e.g. spherocytosis, glucose-6-phosphate dehydrogenase deficiency.
- Extravascular blood, e.g. cephalhaematoma, swallowed blood.
- Increased enterohepatic circulation, e.g. pyloric stenosis, meconium plug.
- Metabolic, e.g. galactosaemia.
- Decreased conjugation, e.g. Crigler–Najjar (rare).
- Sepsis.
- Hypothyroidism.

2. *Conjugated hyperbilirubinaemia*

- Bile duct obstruction, e.g. biliary atresia, choledochal cyst.
- Biliary hypoplasia, e.g. Alagille's syndrome or non-syndromic.
- Neonatal hepatitis. In many conditions, cholestasis represents a final common pathway of injury:
  (a) Metabolic. Galactosaemia, α_1-antitrypsin deficiency, cystic fibrosis,
  (b) Infective. Bacterial infection, congenital viral infection, hepatitis B.
  (c) Endocrine. Hypothyroidism, hypopituitarism.
  (d) Miscellaneous, e.g. chromosomal disorders, parenteral nutrition, lipid storage disorders, idiopathic.

History

Past obstetric history and parental ethnic origins and blood groups should be established, and enquiry made about consanguinity. Illnesses during pregnancy, gestational age and mode of delivery should be noted. The baby may have feeding problems, failure to pass meconium, vomiting, fever or pyrexia. A specific enquiry about the colour of the baby's stool and urine should always be made.

Examination

Weight, length and head circumference should be plotted, and any dysmorphic features noted. Examination may reveal pallor, bruising, purpura or hepatosplenomegaly. Specimens of stool and urine should be viewed by the clinician. More specific diagnostic clues may be present, e.g. cataracts in galactosaemia, cystic mass if there is a choledochal cyst.

Investigation

- FBC, film, reticulocyte count, blood group and save, Coombs' test, maternal haemolysins. If there is cholestasis, a coagulation screen is essential.
- Serum bilirubin, conjugated and unconjugated measurement. Liver function tests, glucose, U&E, acid–base measurement.
- Urine for presence of bile, and for bacterial and viral culture, organic and amino acids and reducing sugars.
- Serum for antibodies to CMV, toxoplasmosis, rubella, etc.
- Serum for α_1-antitrypsin phenotype (total serum level is unreliable), galactose-1-phosphate uridyl transferase activity, amino acids.
- Blood culture.
- Sweat test, plasma immunoreactive trypsin or DNA studies to exclude cystic fibrosis.

- Thyroid function tests.
- Chromosomes.
- DISIDA scan after pretreatment with phenobarbitone will determine bile duct patency. Cerebral and abdominal ultrasound, skeletal survey may be of value.
- Tissue diagnosis – liver biopsy. Occasionally bone marrow or skin biopsy is needed to exclude storage disorders.

Management

In unconjugated hyperbilirubinaemia, the cause of the jaundice should be treated if appropriate, e.g. septicaemia. Unconjugated bilirubin may readily dissociate from albumin and diffuse into the CNS, causing kernicterus, and this is more likely to occur in those with high bilirubin levels, or at lower levels if the neonate is sick and acidotic.

1. Phototherapy. Unconjugated bilirubin is isomerized by exposure to light, and converted into a more water-soluble form, which can then be excreted in bile. Phototherapy is usually considered at bilirubin levels of 250 µmol/l and above in term babies, but at lower levels in the sick or preterm infant. Additional fluids should be given, and the eyes covered.

2. Exchange transfusion. This may be needed for haemolytic disease. A rate of rise of bilirubin levels above 10 µmol/l/h or a positive Coombs' test suggests that exchange transfusion will be needed.

3. Conjugated hyperbilirubinaemia. Immediate priority should be given to the treatment of life-threatening conditions, e.g. sepsis, coagulopathy. The cause of the cholestasis should then be determined with urgency (see investigations above), because biliary atresia requires urgent surgery before 60 days of life if irreversible liver damage is to be avoided. The Kasai procedure achieves biliary drainage by formation of a Roux-en-Y loop of jejunum sutured into the porta hepatis. Other causes of cholestasis may need specific treatment, e.g. surgical excision of a choledochal cyst, and supportive therapy such as coagulation factors, fat-soluble vitamins, added medium-chain triglycerides in feeds.

Further reading

Dodd KL. Neonatal jaundice – a lighter touch. *Archives of Disease in Childhood,* 1993; **68:** 529–32.

Hussein M, Howard ER, Mieli-Vergani G. Jaundice at 14 days of age: exclude biliary atresia. *Archives of Disease in Childhood,* 1991; **66:** 1177–9.

Mowat AP. *Liver Disorders in Childhood,* 2nd edn. London: Butterworths, 1987.

Related topics of interest

Jaundice (p. 195)
Liver disease (p. 208)

NEONATAL RESPIRATORY DISTRESS

Respiratory distress is characterized by tachypnoea (respiratory rate > 60 per minute), expiratory grunting and recession. Cyanosis without oxygen is common. About 2% of all babies and 20% of those under 2500 g have breathing difficulties in the neonatal period. The commonest pulmonary causes of respiratory distress are respiratory distress syndrome (RDS), transient tachypnoea of the newborn, streptococcal pneumonia and meconium aspiration. RDS is due to surfactant deficiency and affects about 7000 UK babies each year. Improvements in neonatal intensive care, including the use of surfactant replacement therapy, have significantly reduced the mortality. The differential diagnosis of a cyanosed baby includes lung disease, cyanotic heart disease and persistent fetal circulation.

Problems

- Hypoxia.
- Acidosis.
- Pneumothorax.
- Bronchopulmonary dysplasia (BPD).
- Sequelae of prolonged ventilation, e.g. airway stenosis.

Differential diagnosis

- Respiratory distress syndrome (hyaline membrane disease).
- Pneumonia (particularly group B streptococcus).
- Transient tachypnoea of the newborn.
- Meconium aspiration.
- Pneumothorax.
- Congenital heart disease (cyanosis, heart failure).
- Diaphragmatic hernia.
- Increased intra-abdominal pressure, e.g. post-repair of gastroschisis.
- Upper airway obstruction, e.g. choanal atresia, Pierre Robin syndrome.
- Cystic lung malformations, e.g. cystic adenomatoid malformation.
- Pleural effusion, e.g. severe rhesus disease.
- Metabolic disorders.
- Myopathies, e.g. Werdnig–Hoffmann, myotonic dystrophy.

Respiratory distress syndrome (RDS)

Clinical features

Predisposing factors include prematurity, hypoxia, hypothermia and maternal diabetes. Boys are more severely affected than girls, and Negro babies are less severely affected than Caucasian babies. Respiratory distress develops within 4 hours of birth and progressively worsens over the next 24–48 hours. Symptoms begin to resolve after 36–48 hours as surfactant synthesis increases, and this is usually associated with a diuresis. Corticosteroids reduce the incidence and severity of RDS when given to mothers in preterm labour. Growth retardation, prolonged rupture of membranes and maternal pre-eclamptic toxaemia advance fetal maturation and may reduce the severity of RDS. The chest radiograph in RDS typically shows granular opacity (ground-glass appearance) with air bronchograms.

Management

1. Supportive care. Management is aimed at supportive care with supplemental oxygen with or without ventilation until spontaneous recovery occurs. Worsening hypoxia, hypercarbia, acidosis and apnoea are indications for ventilation. In mild RDS nasogastric feeding may be possible, but most need intravenous fluids in the acute phase.

2. Surfactant. Pulmonary surfactant is a complex mixture of lipids, proteins and carbohydrate. It is synthesized and secreted by type II epithelial cells, and the quantity increases throughout gestation. It has surface tension-lowering properties in the alveoli and deficiency leads to atelectasis at end-expiration. Two synthetic surfactants have been developed, ALEC (artificial lung expanding compound) and Exosurf, which can be given directly down an endotracheal tube. These have been shown to improve oxygenation and reduce ventilation pressures in babies who clinically develop RDS. Large, randomized, multicentre trials in the last few years have shown a 40% reduction in mortality, and a significant reduction in the incidence of pneumothorax with surfactant therapy. It may also reduce the incidence of ischaemic and haemorrhagic cerebral lesions. Survival with severe BPD is reduced but the overall incidence of BPD seems unchanged.

Surfactant has been given prophylactically to babies at risk of RDS, but there is no clear evidence of the benefit of early administration. Adverse effects include problems with administration (e.g. hypoxia if too rapid, blocked endotracheal tube or if not reconstituted properly) and a possible increase in the incidence of pulmonary haemorrhage in infants < 700 g.

Surfactant therapy in other pulmonary diseases may have a transient beneficial effect, e.g. meconium aspiration, pneumonia, congenital diaphragmatic hernia.

3. *Antibiotics.* It is often difficult to distinguish RDS from early streptococcal pneumonia so penicillin is usually given for the first 48 hours until blood cultures and surface swabs are proven to be negative.

Transient tachypnoea of the newborn

Delayed clearance of fetal lung fluid results in respiratory distress which is usually mild (rarely need > 40% oxygen) and settles in 24–48 hours. Predisposing factors include Caesarean or precipitate delivery. Chest radiography shows diffuse streaky shadowing and fluid in the horizontal fissure.

Meconium aspiration

Acute or chronic intrauterine hypoxia causes 15–20% of babies to pass meconium before or during delivery. Aspiration results in airway obstruction, inhibition of surfactant synthesis and a chemical pneumonitis. The chest radiograph shows patchy atelectasis and hyperinflation due to airway obstruction. The risk of pneumothorax is high, and ventilation should be avoided if possible. Aspiration may be prevented by suctioning the oropharynx immediately after birth. If meconium is seen on the cords at direct laryngoscopy the infant should be intubated and the airway suctioned. The use of bronchial lavage is controversial as it may wash meconium further into small airways. If respiratory distress develops, management is supportive, with supplemental oxygen and i.v. fluids. Antibiotics should be given as there is an increased risk of secondary infection. Symptoms usually settle within 2–7 days.

Bronchopulmonary dysplasia (BPD)

An infant who has a need for supplemental oxygen at 28 days and has an abnormal chest radiograph has BPD. It occurs in ~ 5% of all admissions to neonatal units and is commonest in infants < 1500 g. The most common predisposing factor is ventilation for RDS, but it may occur as a result of other conditions, e.g. meconium aspiration, congenital heart disease. Lung fibrosis and cystic changes occur as a result of barotrauma and oxygen toxicity, and signs of right heart strain may develop. Chest radiography shows coarse shadowing, cysts, hyperexpansion and variable cardiomegaly. Management aims to gradually wean the infant off respiratory support, maintain nutrition (increased calorie requirements) and treat infections promptly with antibiotics and physiotherapy. Diuretics may be of benefit, and steroids have been used to decrease oxygen requirements. The use of steroids is now being advocated earlier in the course of RDS to try to achieve earlier extubation and reduce the incidence and severity of BPD. The use of surfactant has reduced the incidence of severe BPD but there remain a few infants who have a long-term oxygen requirement and are managed with home oxygen. Rehospitalization in the first year of life is common because of respiratory infection, and mortality from respiratory failure or cor pulmonale is increased.

Further reading

Wilkinson AR. Surfactant therapy. In: David TJ, ed. *Recent Advances in Paediatrics,* Vol. 11. Edinburgh: Churchill Livingstone, 1993; 53–66.

Related topics of interest

Congenital heart disease – cyanotic (p. 91)
Prematurity (p. 271)

NEONATAL SURGERY

Disorders of the gastrointestinal tract are common causes of referral for neonatal surgery. Congenital abnormalities may be detected antenatally or present early in the neonatal period. Necrotizing enterocolitis (NEC) is usually managed medically, but surgery is indicated in some cases. Pyloric stenosis, inguinal hernias and intussusception may present in early infancy. Other congenital abnormalities which may require early surgical intervention include congenital heart defects, cleft lip and palate, choanal atresia, congenital lung malformations (e.g. congenital lobar emphysema), urological problems (e.g. posterior urethral valves), sacrococcygeal tumours and biliary atresia, and some of these are discussed elsewhere.

Presentation of gastrointestinal (GI) tract abnormalities in the neonatal period

- Antenatal diagnosis (gastroschisis, diaphragmatic hernia).
- Polyhydramnios (oesophageal atresia).
- Respiratory distress (diaphragmatic hernia).
- Vomiting (intestinal atresia, malrotation, pyloric stenosis, NEC).
- Constipation (meconium ileus, Hirschsprung's disease, anorectal abnormalities).
- Abdominal pain (intussusception, strangulated inguinal hernia).
- Abdominal distension and bloody stools (NEC).

Necrotizing enterocolitis (NEC)

NEC may affect any part of the GI tract, but the commonest sites are terminal ileum, caecum and ascending colon. Damage to the normally impermeable mucosal barrier (hypoxia, ischaemia, immature mucosa) allows organisms, endotoxins and gas to escape from the lumen into the gut wall. NEC is often associated with a septicaemia, but no one organism has been implicated. Nearly all affected infants have received oral feeds, which may act as a substrate for bacterial proliferation. There is an increased risk of NEC with hyperosmolar feeds and when the feed volume is rapidly increased. Breast milk has a protective effect. Coagulation necrosis of the mucosa with microthrombi leads to patchy ulceration, oedema and haemorrhage. This may progress to transmural necrosis and perforation.

Clinical features

NEC can affect term or preterm neonates but is particularly common in those <1500 g. Predisposing factors include hypoxia, hypotension, umbilical catheterization, intrauterine growth retardation, patent ductus arteriosus and exchange transfusion. It can occur at any stage during a baby's stay in a special care baby unit but is most common in the first week of life. Presentation may be non-specific, with lethargy, apnoeas, temperature instability or shock and DIC.

More specific features include abdominal distension, bile-stained vomiting and bloody stools. Bowel sounds are absent, the abdomen is tender and there may be palpable distended bowel loops.

Investigation

- FBC and clotting studies.
- U&E.
- Full septic screen (blood, urine, CSF, stool cultures).
- Abdominal radiography. Dilated bowel loops, thickened gut wall, fluid levels, intramural gas (pneumatosis intestinalis), gas in the biliary tree, evidence of perforation.

Management

1. *Medical.* Most settle with broad-spectrum i.v. antibiotics (including metronidazole) and stopping feeds. Parenteral nutrition should be continued for at least 7 days before feeds are slowly reintroduced. Cardiorespiratory support may be needed and metabolic homeostasis should be maintained. Umbilical catheters should be removed.

2. *Surgery.* Indications for surgery are continued deterioration despite medical treatment, persistent bowel obstruction and perforation. Some perforations can be managed conservatively if the infant is too unstable for surgery.

Outcome

The majority recover fully, but recurrence occurs in ~ 10% of cases, usually within 1 month. Mortality is associated with DIC and septicaemia. Longer term sequelae include strictures, short gut syndrome and fistulae.

Diaphragmatic hernia

Congenital diaphragmatic hernia occurs in about 1 in 2500 live births and is more common on the left side. The infant presents with respiratory distress, usually on the first day of life. Examination reveals a scaphoid abdomen, displacement of the cardiac impulse, and reduced air entry with or without bowel sounds on the affected side of the chest. The degree of pulmonary hypoplasia ranges from mild to severe. Smaller defects, particularly on the right side, may present later and may be confused radiologically with cystic lung lesions (e.g. cystic adenomatoid malformation). Diagnosis is usually made on chest or abdominal radiographs, but can be confirmed by barium swallow. Antenatal diagnosis is becoming increasingly common. Resuscitation with bag and mask ventilation should be avoided as this will cause gastric distension and worsening of respiratory distress. Most affected children are now electively paralysed, ventilated and stabilized for at least 12 hours before surgery. The bowel is decompressed with a nasogastric tube. High ventilation pressures may be needed and there is a risk of pneumothorax. Pulmonary hypertension is treated with hyperventilation and pulmonary vasodilators (e.g. tolazoline). The optimum time for surgery is controversial, but survival is improved if hypoxia and acidosis have been corrected. Survival rates are now better than 50%. The outcome is dependent on the degree of pulmonary hypoplasia and the presence of associated abnormalities. The pulmonary reserve is reduced in the first 1–2 years of life, but the majority of surviving children achieve virtually normal lung function in later childhood.

Oesophageal atresia and tracheo-oesophageal fistula

The majority of cases (> 85%) of oesophageal atresia are associated with a tracheo-oesophageal fistula (TOF). The incidence is 1 in 3000 live births and it is associated with polyhydramnios, which may precipitate premature labour. Affected infants typically have frothy secretions and choke on the first feed. Oesophageal atresia should be excluded in all infants born to mothers with polyhydramnios, by passing a nasogastric tube into the stomach soon after birth. The catheter will not pass more than 8–10 cm from the lips in the presence of oesophageal atresia and the position of the tube can be confirmed radiologically. Air in the stomach confirms the presence of a TOF. The oesophageal pouch is drained with a Replogle tube and all fluids are given intravenously. If there is evidence of aspiration, antibiotics and physiotherapy should be commenced.

Surgery is carried out as soon as the baby is stable. Primary end-to-end anastomosis of the oesophagus is usually possible, with division of the fistula. Contrast studies are performed post-operatively to ensure integrity of the anastomosis before commencing oral feeds. Gastro-oesophageal reflux is common in these children and may require surgical treatment. The brassy TOF cough may persist for several years and is due to a degree of tracheomalacia. Anastomotic stricture may occur.

Gastroschisis and exomphalos

Exomphalos is persistence of the gut herniation which normally occurs between the sixth and 14th weeks of gestation. There may be a large abdominal wall defect covered by a sac of amnion and peritoneum (exomphalos major) or a herniation of abdominal contents into the umbilical cord (exomphalos minor). The incidence is about 1 in 10 000 live births. Exomphalos major has a high incidence (~ 40%) of associated abnormalities (e.g. chromosomal abnormalities, cardiac defects, anencephaly, Beckwith–Wiedemann syndrome). Gastroschisis is evisceration of small and large bowel through a defect of the anterior abdominal wall to the right of the umbilicus. The incidence is about 1 in 6000 live births. The aetiology is unknown. The bowel is often thickened, matted together, and may be shortened. Associated abnormalities are rare. The majority of cases of both gastroschisis and exomphalos are now diagnosed antenatally. Caesarean section does not improve the outcome and these infants can be delivered vaginally. The lesion should be covered with a plastic bag or cling-film to minimize fluid and heat loss, and the bowel decompressed with a nasogastric tube. It may be possible to return the gut to the abdominal cavity as a primary surgical procedure, but larger defects require staged reduction over several days or weeks using a silastic silo. Bowel resection for ischaemic damage is more common with gastroschisis. Intravenous fluids and nutrition are continued until enteral feeding is established post-operatively, which may take weeks or months. Post-operative complications include sepsis, respiratory embarrassment, hypoproteinaemia, renal vein thrombosis and short gut syndrome.

Further reading

Davies CF, Young DG. Gastroenterology. Part IV: Congenital defects and surgical problems. In: Roberton NRC, ed. *Textbook of Neonatology*. 2nd edn. Edinburgh: Churchill Livingstone, 1992: 655–86.

Related topics of interest

Neonatal respiratory distress (p. 235)
Vomiting (p. 328)

NEPHROTIC SYNDROME

The nephrotic syndrome occurs when heavy proteinuria (> 0.05 g/kg/24 h) results in hypoproteinaemia (albumin < 25 g/l) and oedema. It is an uncommon condition, with a prevalence of 6 per 100 000 children and with a peak incidence at age 1–5 years. Boys are more commonly affected than girls (2.5:1). In most cases (>90%) the cause is unknown (idiopathic or minimal change nephrotic syndrome) and the proteinuria is steroid responsive. There is thought to be an immunological basis and there is an association with atopy. The prognosis is good with < 10% progressing to chronic renal failure. Genetic predisposition is associated with HLA-DRW7 and - DR3 (sixfold increased risk) and there is an increased risk in siblings of an affected child. Less commonly the nephrotic syndrome occurs secondary to another disorder and is not steroid responsive. In these children the prognosis is generally poor, more than 30% progressing to renal failure.

Problems

- Hypovolaemia.
- Infection.
- Thrombosis.
- Acute renal failure.
- Complications of prolonged steroid treatment, e.g. poor growth, obesity, hypertension.

Aetiology and pathology

1. *Idiopathic (minimal change, MCNS).*

2. *Secondary to:*
- Henoch–Schönlein purpura.
- Glomerulonephritis.
- Malaria.
- Renal vein thrombosis.
- Toxins, e.g. lead, gold, snake venom.
- Allergy, e.g. bee sting, immunization.
- Miscellaneous conditions, e.g. amyloid, sickle cell anaemia, systemic lypus erythematosus (SLE).

3. *Congenital.* The light microscopic findings in MCNS are unremarkable, and electron microscopy shows fusion of the epithelial cell foot processes. In the steroid-unresponsive group the underlying pathology is usually focal segmental glomerulosclerosis (less commonly membranoproliferative glomerulonephritis).

Clinical features

Oedema is the usual presenting feature, typically facial and periorbital. There may be associated peripheral oedema, ascites and pleural effusions. The child is often irritable, with a poor appetite, and may have diarrhoea and/or vomiting. Symptoms and signs of hypovolaemia, including cold peripheries and abdominal pain (due to splanchnic underperfusion), are very important as hypovolaemia can lead to peripheral ischaemia and infarction, resulting in loss of digits and even limbs. Hypertension may be a feature if the nephrotic syndrome is secondary to glomerulonephritis.

Congenital nephrotic syndrome is very rare, and presents at birth with placental oedema, or develops in the first few months of life. It is an autosomal recessive condition, more common in Scandinavia.

Investigation

- *Urinalysis.* Proteinuria. Some centres determine the protein selectivity index, the proteinuria of MCNS being highly selective (i.e. small proteins only, e.g. albumin) and of focal segmental glomerulosclerosis being poorly selective. However steroid responsiveness is the best indicator of MCNS. Haematuria or granular casts indicate an underlying glomerulonephritis.
- *Full blood count and haematocrit.* A useful guide to hypovolaemia.
- *U&Es, creatinine, albumin.* Urea may be raised due to hypovolaemia but creatinine is usually normal in MCNS. Hyponatraemia may occur.
- *Calcium and magnesium.* Levels reduced in parallel with albumin (ionized calcium normal).
- *Complement screen and immunoglobulins.* IgG and serum C3 are reduced. In 25% there is raised IgE.
- *Hyperlipidaemia* is common.

Management

1. Correct hypovolaemia. Plasma expansion with salt-poor albumin and cautious use of diuretics (may induce hypovolaemia). Consider monitoring central venous pressure in the acute phase. Acute renal failure may occur as a result of hypovolaemia or the underlying condition. Strict fluid input/output monitoring including daily weighing is essential. A low-sodium diet and bed rest may help to reduce oedema.

2. Induce remission. In MCNS, remission will occur spontaneously given time (months or years), but it is steroid responsive and should be treated with prednisolone. Before steroid therapy as many as 30% of children with MCNS died, mainly because of overwhelming infection. Many different steroid regimens are used, e.g. prednisolone 2 mg/kg/day until protein-free or for 4 weeks, then tailing off over 6 weeks. Relapses are common (> 75% have at least one relapse) and are often associated with infections or exacerbations of asthma/eczema. These are treated with further courses of prednisolone. Frequent relapses may warrant the use of alternate-day prednisolone for up to 6 months to prevent recurrences. If relapses occur on prednisolone or the side-effects of steroids become unacceptable, levamisole, cyclophosphamide or cyclosporin A may be considered. Renal biopsy is not performed routinely and is only indicated in those who fail to respond to steroids or in those with poor prognostic features at diagnosis (haematuria, hypertension, raised creatinine, age <1 year or > 10 years).

3. Prevent infections. Nephrotic patients are immunocompromised because of low IgG levels, impaired lymphocyte function, protein malnutrition and immunosuppressive drugs. Prophylactic penicillin should be given to oedematous patients to prevent pneumococcal peritonitis. Fever will be masked by steroid treatment and abdominal pain may be the only indication of peritonitis.

4. Prevent thrombosis. The increased clotting tendency is due to elevated clotting factors and enhanced platelet aggregability, and may be exacerbated by hypovolaemia. Renal vein thrombosis may further complicate the nephrotic syndrome.

Further reading

Consensus Statement on Management and Audit Potential for Steroid Responsive Nephrotic Syndrome. Report of a Workshop by the British Association for Paediatric Nephrology and Research Unit, Royal College of Physicians. *Archives of Disease in Childhood*, 1994; **70:** 151–7.

Watson AR. Nephrotic syndrome. In: Campbell AGM, McIntosh N, eds. *Forfar and Arneil's Textbook of Paediatrics.* 4th edn. Edinburgh: Churchill Livingstone, 1992: 1057–61.

Watson AR, Coleman JE. Dietary management in nephrotic syndrome. *Archives of Disease in Childhood*, 1993; **69:** 179–80.

Related topics of interest

Acute renal failure (p. 10)
Chronic renal failure (p. 75)
Glomerulonephritis (p. 146)

NEUROCUTANEOUS SYNDROMES

These disorders are syndromes of genetic or developmental anomalies which involve structures of ectodermal origin, i.e. peripheral and central nervous system, skin and eye. Only some of these conditions are discussed below.

Problems

- Neurological.
- Cosmetic.
- Visceral.
- Developmental delay.
- Visual disturbance.

Tuberous sclerosis (TS)

The classical features of this autosomal dominantly inherited condition are mental retardation, seizures and facial angiofibromas. However, it has a wide range of clinical features, and a high mutation rate accounting for up to two-thirds of cases. Birth prevalence is thought to be as high as 1 in 6000. A gene locus has been found on the long arm of chromosome 9, but also on other sites, e.g. 11q, and this genetic heterogeneity means that antenatal diagnosis is not yet possible.

Skin lesions include hypomelanic 'ash leaf'-shaped macules which fluoresce under UV light (Wood's lamp) and may be present from birth. The fibrous forehead plaque may also appear early.

Adenoma sebaceum rarely appears before the age of 2–3 years and is a papular acneiform eruption in a butterfly distribution over the face, sparing the upper lip. Periungual fibromata of fingers or toes are rarely seen before puberty, but may be the sole sign of TS. Shagreen patches are raised, roughened areas of skin over the lumbosacral area. Neurological features include mental retardation (50%) and seizures, commonly infantile spasms. Psychotic and autistic behaviour is common. Intracranial lesions occur, e.g. subependymal glial nodules, cortical tubers and giant cell astrocytomas (rare but can cause hydrocephalus by obstruction to the third ventricular outflow).

Cardiac rhabdomyoma may be an important prenatal marker, and may present with obstructive cardiac symptoms but then regress later. Renal involvement occurs in 80% of cases, as polycystic kidneys or angiolipomata.

Retinal phakomata may be a feature, normally symptomless, appearing as a fleshy, raised lesion, often around the optic disc.

Management of TS should involve effective control of seizures, and active surveillance for the recognized complications of the disease. Argon laser therapy has been successful in the treatment of facial angiofibromas.

Sturge–Weber syndrome

The main features of this rare and sporadically occurring syndrome are leptomeningeal angiomatosis (all cases), facial angiomatous naevus of the ophthalmic division of the trigeminal nerve (85% of cases) and choroidal angioma (40% of cases). The leptomeningeal angioma is common in the parieto-occipital area, and is bilateral in 15% of cases. The facial port-wine stain (angiomatous naevus) is usually ipsilateral to the leptomeningeal angioma.

Clinical features are seizures (often partial motor seizures in early infancy), mental retardation, glaucoma and hemiplegia. Characteristic appearances on skull radiographs are of 'rail-road' track calcification after the first year of life, but MRI and CT scanning will reveal changes earlier.

Management involves the control of seizures, regular ophthalmological assessment and laser treatment of skin lesions. Neurosurgery may be suitable in selected cases.

Neurofibromatosis (NF)

This is the most common of the neurocutaneous syndromes, and can be separated into type 1 neurofibromatosis (classical, von Recklinghausen), assigned to chromosome 17, and type 2 (central) neurofibromatosis, assigned to chromosome 22. Both are autosomal dominant disorders.

1. Type 1 neurofibromatosis (NF1). Two or more of the following diagnostic criteria must be present for diagnosis:

- Six or more *café au lait* macules (over 5 mm diameter in prepubertal children and over 15 mm in post-pubertal children). These often appear in the first year of life, and are nearly always present by 6 years of age.
- Two or more neurofibromas of any type, or one plexiform neurofibroma.
- Axillary or inguinal freckling (tends to follow the appearance of *café au lait* lesions).
- Optic glioma.
- Two or more Lisch nodules.
- A distinctive osseous lesion such as sphenoid dysplasia or pseudoarthrosis.
- A first-degree relative with NF1.

Complications include malignant change in the skin lesions, especially the diffuse subcutaneous plexiform tumours. CNS malignancy (optic nerve pathway gliomas), phaeochromocytoma, renal artery stenosis, spinal (dumb-bell) tumours, gastrointestinal neurofibromas and epilepsy are all rare but serious complications.

2. *Type 2 neurofibromatosis (NF2).* It is unusual for NF2 to present in the paediatric age group (average age of onset is 20 years), and new mutations are common. Symptoms occur from bilateral acoustic neuromas, cranial meningiomas and spinal tumours. Skin signs may be minimal.

Incontinentia pigmenti

This is an X-linked dominant disorder, usually lethal in males. A triphasic rash (vesicular, verrucous, hyperpigmented) tends to fade altogether by adulthood. Seizures, mental retardation, spasticity and ocular abnormalities are associated features.

Klippel–Trenaunay–Weber syndrome

Complex angiomatous malformations occur over one or more limbs, associated with hypertrophy of soft tissue and bone. Cerebral angiomatous malformations may occur.

Further reading

De Sousa C. The neurocutaneous syndromes. *Current Paediatrics*, 1992; **2(2):** 86–9.

Evans G. Neurofibromatosis. *Current Paediatrics*, 1992; **2(2):** 83–5.

Webb DW, Osborne JP. Tuberous sclerosis. In: David TJ, ed. *Recent Advances in Paediatrics*, Vol. 11. Edinburgh: Churchill Livingstone, 1993: 147–60.

Related topics of interest

Developmental delay and regression (p. 115)
Hydrocephalus (p. 174)
Rashes and naevi (p. 285)
Seizures (p. 288)

NUTRITION

Nutritional care has an impact on all infants and children, whether well or sick. For the seriously ill child, nutritional support is now available in many different forms.

Problems

- Protein–energy malnutrition.
- Vitamin and mineral deficiencies.
- Food intolerance.
- Specific dietary regimens, e.g. inborn errors of metabolism.
- Nutrition of the preterm infant.
- Long-term outcome of childhood nutrition.

Assessment of nutritional status

The history may provide information about an underlying disorder. If available, a full analysis of dietary intake is invaluable. On a practical basis, clinical, anthropometric or laboratory assessments can be made, whereas body composition analysis allows more detailed analysis for research purposes.

Clinical assessment may show wasting, apathy, bony deformity or signs of specific vitamin deficiencies, e.g. perifollicular haemorrhages in vitamin C deficiency. Anthropometric indices are weight, length, head circumference, mid-upper arm circumference and skinfold thickness on the triceps and subscapular region. Biochemical and haematological indices provide information about specific nutrients.

Breast-feeding

Breast milk provides complete nutrition for the healthy term infant, and breast-feeding should be promoted by all health-care workers. It is an uncontaminated milk source, containing secretory IgA and lactoferrin. Breast-feeding reduces the risk of gastrointestinal and respiratory infection. Its unique composition of vitamins, fats, carbohydrate, protein, minerals and enzymes ensures optimum absorption and trophic effect on the gut. Mother–infant bonding is enhanced, and suckling promotes oromotor function.

Alternative milk feeds

For mothers who are unable to breast feed or do not wish to, a modified cow's milk formula should be chosen. The casein – whey ratio should mimic the composition of breast milk, i.e. 40:60, and numerous suitable commercial brands are available, with little variation between them. Casein-based milks claim to provide greater satisfaction for a hungry baby, but there is little evidence to support this. Unmodified cows' or goats' milks are unsuitable alternatives because of high solute loads, vitamin and iron deficiencies. Although recommended volumes in young infants are 150 ml/kg/day (105 kcal/kg/day), there is enormous variation, and evidence of a sufficient milk intake should be guided by the infant's growth and state of well-being.

Preterm formula milks

These formula milks provide a greater density of nutrients, aiming to achieve intrauterine growth rates without feeding large volumes. Studies run by Lucas at the Dunn nutrition unit show that infants fed such milks achieve better growth, motor and mental performance in early childhood than those on banked breast or standard formula milk. However, Lucas also reports that human milk may confer advantages in terms of improved IQ and protection against necrotizing enterocolitis. Human milk fortifiers (prepared fractions of human milk) have now been developed to enrich human milk for use in the very low birth weight infant. Short- and long-term effects of nutrition in very low birth weight infants are the subject of current investigation and debate.

Cow's milk

The introduction of 'doorstep' cows' milk should be deferred until the age of 1 year, as the iron and vitamin content is too low, and subclinical gastrointestinal bleeding may develop. Iron deficiency has an adverse effect on immunological, neuromuscular and cognitive function. Studies suggest that iron deficiency occurs in 25–50% of infants and young children. The commercial 'follow-on' milks have emerged, to provide a milk with a higher iron, protein and vitamin concentration than standard formula milks after 6 months of age.

Enteral nutrition

Nutritional support via the enteral route is the method of choice if small bowel function is intact. Indications include prematurity, anatomical problems of the oropharynx, chronic cardiorespiratory, renal or liver disease, severe anorexia nervosa and primary gastrointestinal disease such as short-gut disease or Crohn's disease. Feeding may be via a nasogastric or gastrostomy tube (which may be placed percutaneously), although rarely transpyloric feeding is more suitable. Specialized formula feeds vary in their calorific content and osmolality, and an elemental formula may be indicated for induction of remission in Crohn's disease, or for use in malabsorption syndromes.

Total parenteral nutrition (TPN)

This should be instituted when the gut cannot be used, e.g. in short-gut syndrome, inflammatory bowel disease, hyaline membrane disease, necrotizing enterocolitis, chronic intestinal pseudo-obstruction and numerous other situations. Circulatory access may be via peripheral veins or via a centrally placed catheter. TPN provides a balanced preparation of amino acids, lipid, carbohydrate, vitamins, electrolytes and trace elements, ideally prepared in a pharmacy department. Close clinical, microbiological and biochemical monitoring is essential. Complications arise from the route of access (extravasation, catheter sepsis or occlusion) and from biochemical disturbance such as hyperglycaemia, hyperlipidaemia and trace metal deficiencies. With long-term bowel rest, small bowel intestinal atrophy and pancreatic atrophy may occur.

Further reading

Dalzell AM, Dodge JA. Enteral nutrition. *Current Paediatrics*, 1992; **2(3):**168–71.

Wharton B. Milk for babies and children: no ordinary cow's milk before one year. *BMJ*, 1990; **301:** 774–5.

Related topics of interest

Chronic diarrhoea (p. 72)
Failure to thrive (p. 133)
Prematurity (p. 271)

OSTEOMYELITIS

Acute osteomyelitis can occur at any age, the commonest causative organism being *Staphylococcus aureus*. Infection is spread to bone, either haematogenously from a distant focus (e.g. skin sepsis) or directly from a penetrating injury. Long bone metaphyses are most commonly affected, but any bone can be involved including craniofacial bones, ribs and vertebral bodies. It is particularly severe in the neonatal period, when it is frequently multifocal. Chronic osteomyelitis occurs if the acute infection is inadequately treated, or if the organism is of low virulence.

Causative organisms

- *Staphylococcus aureus (> 70%).*
- Group A and B streptococcus.
- *Haemophilus influenzae.*
- *Streptococcus pneumoniae.*
- Enteric and Gram-negative organisms.
- Group B haemolytic streptococcus. Increasingly important in neonates
- *Salmonella.* Associated with sickle cell anaemia
- *Candida albicans.* May cause acute or chronic osteomyelitis in immunosuppressed patients or after prolonged intravenous feeding.
- *Mycobacterium tuberculosis.* Usually chronic osteomyelitis.
- Other rare causes of chronic osteomyelitis, e.g. actinomycosis.

Clinical features

Onset of acute osteomyelitis is usually sudden. The distant focus from which infection has spread may be obvious (e.g. umbilical cord or skin sepsis), but it is frequently clinically inapparent. The child rapidly becomes toxic, febrile and unwell, and is reluctant to move. Vomiting and headache are often prominent features in older children, and neonates may present with circulatory collapse. The affected limb is swollen, red and painful, and the adjacent joint may contain a sympathetic effusion. Spread of infection into the joint occurs in the neonatal period as vascular connections still exist, but in older children spread is normally prevented by the epiphyseal growth plate.

Chronic osteomyelitis usually presents with a grumbling fever associated with bone pain and/or swelling, which must be distinguished from neoplastic conditions. This usually necessitates a biopsy of the affected bone.

Differential diagnosis of bone pain

1. *Swelling*

- Osteomyelitis (acute/chronic).

- Fracture.
- Infantile cortical hyperostosis (Caffey's disease.)
- Advanced osteosarcoma.

2. *No swelling* .
- Osteomyelitis.
- Osteoid osteoma.
- Osteosarcoma.
- Ewing's sarcoma.
- Neuroblastoma.
- Leukaemia.

Investigation

- FBC. Neutrophilia.
- Blood cultures. Positive in more than 50% of cases.
- ESR. Raised.
- Radiology. No bony changes are seen on radiographs for 5–10 days, although there may be some soft-tissue swelling. Spotty rarefaction and subperiosteal new bone formation are the first radiological changes.
- Isotope bone scan. A technetium scan may detect an area of osteomyelitis before radiological changes are seen.

Management

1. *Antibiotics.* Early diagnosis and treatment with high-dose i.v. flucloxacillin with either gentamicin, fucidin or ampicillin is vital for effective treatment. Antibiotic choice will be guided by results of culture and sensitivity, and cephalosporins may be needed for ampicillin-resistant *Haemophilus influenzae*. Antibiotics are continued for at least 6 weeks, and for at least 1 week should be given intravenously. Chronic osteomyelitis is treated with a prolonged course of antibiotics (6–12 months).

2. *Surgical drainage.* Pus may collect in the medullary cavity, subperiosteally or subcutaneously and may require drainage. In joints in which the long bone metaphysis is intracapsular, e.g. the hip, pus can rupture into the joint, causing a septic arthritis.

3. *Pain relief.* Pain may be very severe and treatment includes analgesics, splinting of the limb and release of pus that is under tension.

4. *General supportive measures.* These children are often extremely unwell. Vomiting should be managed with intravenous fluids and nasogastric aspiration. Anaemia is common, requiring blood transfusion in the acute phase and iron supplements during recovery.

5. *Physiotherapy* is employed after the acute phase to prevent contractures and strengthen muscles.

Outcome

If diagnosed and treated early the prognosis is good. However, extensive destruction of the bony cortex can lead to pathological fractures and inequality of limb length. Osteomyelitis of the upper femoral metaphysis can result in severe bone and cartilage destruction if detected late when pus is already present in the hip joint. This leads to major problems with subsequent mobility. Osteomyelitis of a vertebral body can be complicated by paravertebral abscess and pyogenic meningitis. Incomplete treatment of acute osteomyelitis may result in chronic infection and discharging sinuses.

Further reading

Kliegman RM, Feigin RD. Osteomyelitis. In: Behrman RE, ed. *Nelson Textbook of Paediatrics*, 14th edn. Philadelphia: WB Saunders, 1992; 691–4.

Related topic of interest

Limping and joint disorders (p. 205)

PERIPHERAL NEUROPATHY

Peripheral nerve disorders may affect sensory and/or motor nerves, involving one nerve (mononeuropathy) or causing a more generalized neuropathy (polyneuropathy). A neuropathy can present acutely (e.g. foot drop, facial palsy) or more chronically with muscle wasting and weakness. Sensory nerve involvement results in paraesthesia or anaesthesia, typically of the hands and feet. Neurophysiology shows slowed nerve conduction velocity and muscle biopsy may show denervation.

Aetiology

- *Trauma.* Birth injuries, fractures.
- *Hereditary degenerative.* Motor and sensory neuropathies, peroneal muscular atrophy (Charcot–Marie–Tooth).
- *Metabolic.* Diabetic (rare in childhood), Refsum's disease, metachromatic leucodystrophy.
- *Toxins.* Vincristine, lead, solvent abuse.
- *Deficiencies.* B_{12}, thiamine (rarely cause neuropathy in childhood).
- *Parainfectious.* Guillain–Barré syndrome.
- *Idiopathic.*

Guillain–Barré syndrome

This is an uncommon condition which mainly affects older children and adults but has been reported in infants as young as 6 months. A mild preceding febrile illness is common. Typically, there is sudden onset of symmetrical flaccid paralysis affecting the legs first and then ascending to involve the trunk and arms. Sensory loss is often a prominent early feature in children, with paraesthesiae causing severe limb pain and ataxia. Limb weakness and hyporeflexia may be unimpressive initially, leading to difficulties with diagnosis. Meningism and papilloedema are present in about one-third of cases and may cause further diagnostic confusion. Progression of muscle weakness may continue for several days and even weeks. Involvement of respiratory muscles necessitates mechanical ventilation in 10–20% of cases. In severe cases bilateral facial palsy and bulbar palsy occur. Autonomic involvement may cause hypertension and arrhythmias, excessive sweating, disturbed gut function and urinary retention.

The condition is usually self-limiting, with more than 80% making a full recovery, but the convalescent period may be very prolonged.

The underlying pathophysiology is demyelination of the peripheral nerves, which is thought to occur as a result of an autoimmune reaction. Suggested aetiological agents include mycoplasma and viruses, such as coxsackie B and Epstein–Barr virus.

Investigations

- *CSF.* Raised protein (may be normal early in illness). Normal white cell count.

- *Viral cultures and serology, Paul Bunnell test.*
- *Neurophysiology* (may be normal early in illness).
- *Urine for porphyrin screen.* Exclude acute intermittent porphyria.
- *U&E.* Exclude hypokalaemic periodic paralysis.
- *Radiology.* Exclude spinal cause of flaccid paralysis.
- *Toxicology screen*, e.g. exclude lead poisoning.

Management

1. Supportive care. If Guillain–Barré syndrome is suspected, the child should be admitted. A peak flow rate is measured 4-hourly in older children. Transcutaneous oxygen monitoring is useful in children unable to do a peak flow, and arterial blood gas measurements may be necessary. A falling peak flow rate, hypoxia, hypercarbia and exhaustion are indications for ventilation. The blood pressure should be monitored for evidence of autonomic involvement and the child may require catheterization for urinary retention. Regular turning and passive exercises help prevent the development of pressure sores and contractures. Pain relief may be needed for severe muscle pain. Nutrition is important and fluids should be given intravenously because of the risk of aspiration. Paralysis without impairment of awareness can be terrifying and requires reassurance and good communication. Sedation may be indicated.

2. Specific treatments. Steroids have not been shown to improve the outcome. Plasmapheresis may produce a clinical improvement and speed recovery if used early in the acute phase, but should be reserved for the most severely affected patients. Intravenous gammaglobulin has been reported to be helpful but has only been used in small studies.

3. Rehabilitative physiotherapy.

Outcome

Acute mortality from respiratory failure and arrhythmias (including cardiac arrest) is now less than 5%. Over 80% make a complete recovery, but this may take several years. Severity of initial weakness does not correlate with long-term prognosis, but evidence of early improvement (before 3 weeks) is a good prognostic indicator.

Facial palsy

Aetiology

- *Idiopathic* (Bell's palsy).
- *Infections,* e.g. Ramsay Hunt (herpes zoster), infectious

mononucleosis, mastoiditis, Lyme disease (characteristic rash, polyarthritis).

- *Trauma,* e.g. forceps delivery.
- *CNS leukaemia.*
- *CNS tumour,* e.g. pontine glioma.
- *Hypertension.*
- *Guillain–Barré syndrome* – usually bilateral.
- *Myasthenia gravis.*
- *Post-ictal* (Todd's paralysis).
- *Drugs,* e.g. vincristine.

Clinical features

Facial palsy in childhood is usually idiopathic, but may be associated with viral infections or Lyme disease. There is sudden onset of lower motor neurone weakness of one side of the face. Problems include conjunctivitis and corneal abrasion (owing to inability to close the eye), dribbling and slurring of speech. There may be facial pain either at the onset or during recovery. More than 80% will recover spontaneously with some improvement seen within 3 weeks.

Management

1. Careful history and clinical examination. Exclude hypertension and features of an intracranial tumour (headache, vomiting, other neurological signs), CNS leukaemia (bruising, anaemia) and middle ear disease. The majority of cases will be idiopathic and require symptomatic treatment and observation only. Investigations such as FBC, viral titres, Lyme disease serology or brain imaging may be indicated on the basis of clinical findings.

2. Pain relief and prevention of conjunctival irritation. Taping eye closed, artificial tears.

3. Steroids. May be beneficial if given within the first 48 hours, but the evidence to support their use is mixed. FBC to exclude leukaemia is essential before giving steroids.

4. Surgery. A few cases do not resolve and these patients may require plastic surgery to restore facial symmetry and protect the eye.

Lead poisoning

Chronic lead poisoning is a rare cause of peripheral neuropathy. Other features include anorexia, vomiting, abdominal pain, constipation, behavioural problems and failure to thrive.

More severe poisoning may cause an encephalopathy with seizures, drowsiness and coma. Sources of lead include paint, old water pipes, batteries, surma eye make-up used by Asians and pollution. Symptoms are unlikely if the blood lead level is < 80 μg/100 ml. Investigations show a microcytic hypochromic anaemia, basophilic stippling, raised plasma δ-aminolaevulinic acid and free erythrocyte porphyrin, and raised urinary co-proporphyrin. Radiographs may show lead lines at the end of long bones. Management consists of identifying and removing the source of lead and the use of chelating agents (e.g. D-penicillamine, EDTA).

Further reading

The Guillain–Barré Study Group. Plasmapheresis and acute Guillain–Barré syndrome. *Neurology,* 1985; **35:** 1096–104.

Related topic of interest

Floppy infant (p. 136)

POISONING

Accidental poisoning is very common, resulting in over 40 000 attendances at casualty each year. It is most common in young children and toddlers, with a peak age incidence at 2.5 years, and boys are more often the culprits than girls. A wide variety of medicines, household products and plants are ingested. Fewer than 30 children a year die as a result of poisoning, but large numbers of children are seen and observed in an attempt to prevent these occasional deaths. Education of parents about the dangers of poisoning and the use of lockable cupboards, childproof containers and blister packaging have all helped to reduce the incidence of serious poisoning. Intentional self-poisoning is most common amongst adolescent girls, and occasionally children are poisoned by an adult as part of the Munchausen by proxy syndrome. Poisoning may also result from alcohol or solvent abuse. There may be no history of poisoning, but if a previously well child presents with unexplained symptoms (e.g. drowsiness, coma, convulsions, respiratory depression) then a diagnosis of poisoning should be considered, and blood and urine should be saved for toxicology.

Management

Establish exactly what and how much has been ingested, and how long ago. Advice can be sought from regional poisons units on the possible toxic effects of various substances, and the availabilty of antidotes or specific removal methods (e.g. alkaline diuresis). The majority of children do not have serious symptoms, but occasionally a child presents with respiratory depression, hypotension or reduced consciousness and requires cardiorespiratory resuscitation and support.

If the substance ingested might be harmful, vomiting should be induced (unless contraindicated) with 15 ml of ipecacuanha syrup given with a drink. The dose of ipecac can be repeated after 20 minutes if there has been no effect. Ninety-five per cent of children will vomit, and they should then be observed for a few hours for side-effects of the poison before being sent home. All symptomatic children and those who have ingested iron, tricyclics or digoxin should be admitted. Indications for gastric lavage are few (e.g. iron, paraquat poisoning). Activated charcoal reduces the absorption of some poisons, but it is almost impossible to get a child to drink it. It may be useful in conjunction with gastric lavage. Full psychiatric and social assessment is essential in cases of intentional self-poisoning.

Contraindications to emesis

- More than 4 hours since ingestion (except aspirin, lomotil, tricyclics).
- Corrosive substances.
- Volatile substances, e.g. turpentine, paraffin (risk of inhalation and chemical pneumonitis).

- Decreased consciousness. Intubate with a cuffed tube and empty the stomach with gastric lavage.

Salicylates

Problems

- Hyperventilation. Deep sighing respirations due to metabolic acidosis.
- Hyperglycaemia.
- Tinnitus.
- Hyperpyrexia, vasodilatation, sweating.
- Vomiting and abdominal pain.
- Dehydration.

Management

1. Emesis should be induced even up to 24 hours post-ingestion as absorption is often delayed.

2. Measure blood salicylate level at 4–6 hours post-ingestion. A level of <40 mg/100 ml requires no treatment. Admit the child if the level is > 40 mg/100 ml, the child is symptomatic or if oil of wintergreen has been ingested (oil of wintergreen contains methylsalicylate and is extremely dangerous – only one teaspoonful can be lethal). If the level is mildly raised (40–65 mg/100 ml) encourage a high fluid intake and commence i.v. fluids if the patient is not drinking well. If the level is high (> 65 mg/100 ml) forced alkaline diuresis is needed. A level of 120 mg/100 ml is usually lethal.

3. Supportive management of severe poisoning. Monitor acid–base status (metabolic acidosis, respiratory alkalosis), blood glucose (mild to moderate hyperglycaemia), U&E and prothrombin time. Tepid sponging may be necessary for hyperpyrexia. Strict fluid balance is essential with the risk of fluid overload from forced alkaline diuresis, and the risk of dehydration from vomiting, hyperpyrexia and hyperglycaemia.

Paracetamol

Problems

- Nausea and vomiting. Lack of early symptoms other than some nausea and vomiting may be deceptive.
- Acute hepatic failure. Damage due to hepatocellular necrosis is maximal at 3–4 days post ingestion and leads to confusion, drowsiness, coagulopathy and jaundice.
- Acute tubular necrosis.

Management

1. *Emesis* should be induced if within 4 hours of ingestion.

2. *Measure blood paracetamol level* at least 4 hours post-ingestion and compare with a nomogram of level against time since ingestion. A level of < 200 μg/ml at 4 hours requires no treatment. Levels above a line joining 200 μg/ml at 4 hours and 30 μg/ml at 15 hours on a semilogarithmic graph require treatment. Patients on liver enzyme-inducing drugs (e.g. carbamazepine) may develop toxicity at lower paracetamol levels.

3. *Methionine or acetylcysteine.* These protect the liver if given within 12 hours of ingestion. Methionine is given orally as four doses over the first 12 hours. Acetylcysteine is given as an i.v. infusion over 20 hours.

4. *Monitor prothrombin time and liver function tests* for 3–4 days for evidence of hepatocellular damage if the paracetamol level was high.

5. *Supportive management of acute liver failure.*

Iron

Problems

- Phase 1 (6–12 hours), gastric irritation. Nausea, vomiting, abdominal pain, gastrointestinal haemorrhage.
- Phase 2 (12–24 hours), quiescent phase. Asymptomatic. Deposition of iron in the liver.
- Phase 3 (onset 16–24 hours), cardiovascular collapse, acute encephalopathy and hepatic failure.
- Phase 4, pyloric scarring and stenosis. If phase 3 is survived.

Management

1. *Emesis.* All patients should be admitted and vomiting induced. As little as 2 g of elemental iron may be fatal.

2. *Desferrioxamine.* Emesis should be followed by gastric lavage with desferrioxamine to chelate the iron. Desferrioxamine by i.v. infusion is indicated if the serum iron level is > 5 mg/l.

3. *Supportive management.*

Further reading

Poisoning. In: Advanced life support group. *Advanced Paediatric Life Support – the Practical Approach.* London: BMJ Publications, 1993; 121–6.

Related topics of interest

Cardiopulmonary resuscitation (p. 52)
Jaundice (p. 195)
Liver disease (p. 208)
Shock (p. 291)

POLYARTHRITIS

Polyarthritis in childhood is most commonly due to an infection or Henoch–Schönlein purpura. Infections may directly involve the joints or may cause a reactive arthritis (e.g. rubella). Chronic inflammatory bowel disease and gastrointestinal infections may be associated with an arthropathy. JCA is relatively uncommon, affecting 1 in 5000 children, but it is an important cause of chronic handicap. Rheumatic fever, associated with group A β-haemolytic streptococcus infection, is now uncommon in the UK but is a significant cause of morbidity and mortality in developing countries. Trauma usually affects one joint but occasionally presents with involvement of several joints.

Aetiology

- Infection. Bacterial, tuberculous, viral, Lyme disease.
- Henoch–Schönlein purpura.
- Juvenile chronic arthritis.
- Gastrointestinal disorders. Crohn's, ulcerative colitis, shigella dysentery, *Yersinia.*
- Haematological. Haemophilia.
- Rheumatic fever.
- Trauma. Accidental, non-accidental.
- Systemic lupus erythematosus, dermatomyositis, psoriasis.

Juvenile chronic arthritis (JCA)

Problems

- Joint destruction and deformity. Particularly polyarticular JCA.
- Uveitis. Particularly ANA-positive pauciarticular JCA.
- Amyloidosis.
- Side-effects of treatment, e.g. steroids, hydroxychloroquine (retinopathy), D-penicillamine (blood dyscrasias, proteinuria).
- Family and social disruption.

Clinical features

JCA causes pain, swelling and stiffness of one or more joints. Diagnosis depends on three criteria:

- Onset under 16 years.
- Arthritis of one or more joints for a minimum of 3 months.
- Exclusion of other diseases, e.g. SLE, psoriasis, GI tract disease.

Current classification divides JCA into three main groups which have different natural histories. A few children progress from one group to another:

1. Pauciarticular (50%) involving four joints or fewer. The usual age of onset is 1–5 years and it is more common in girls. There are usually no systemic symptoms. If the onset is monoarticular, the knee is most frequently affected. The cervical spine, jaw and the small joints of the hands and feet are rarely affected. Joint disease is rarely progressive and usually resolves by the age of 16 years. Chronic iridocyclitis (uveitis) is a major problem and ophthalmic review is essential (increased risk if ANA positive).

Investigations

- Positive antinuclear antibody (70%).
- Negative rheumatoid factor.
- Association with HLA DRW5/8.

2. Systemic onset (Still's disease). Arthritis may be delayed for weeks or even months, and the major features at the onset are systemic (intermittent high fever, splenomegaly, rash, lymphadenopathy). Onset is usually under 5 years of age and it is equally common in boys and girls. It may take many weeks to make the correct diagnosis, which depends on clinical features and exclusion of infection (viral, bacterial, mycoplasma) and neoplastic disorders (leukaemia, neuroblastoma). Pericarditis is relatively common, and hepatitis and myocarditis are uncommon complications. The arthritis is characteristically relapsing and remitting. Amyloidosis may develop, usually presenting with proteinuria.

Investigations

- ANA and rheumatoid factor negative.
- Raised ESR.
- Raised WBC.
- Raised platelets.
- Anaemia.

3. Polyarticular. Five or more joints are involved, commonly the knees, ankles, small joints of the hands and feet, the jaw and cervical spine. Joint swelling and stiffness may be painful or relatively pain-free. Onset can occur at any time from 1 to 15 years. Joint involvement usually follows a chronic, progressive course. Involvement of the temporomandibular joints can produce micrognathia. Systemic features are minimal but there may be a low-grade fever, weight loss and anaemia. Uveitis is rarely a feature.

Investigations

- ANA usually negative.
- Rheumatoid factor negative.

Juvenile rheumatoid arthritis (positive rheumatoid factor) and juvenile ankylosing spondylitis (anterior uveitis, HLA B27) may present in later childhood and follow a similar course to that in adults.

Management

Effective management depends on a multidisciplinary approach (paediatrician, rheumatologist, GP, orthopaedic surgeon, ophthalmologist, physiotherapist, occupational therapist, social worker, school). Treatment is supportive, to relieve symptoms and maintain function, but is not curative.

1. Analgesics, e.g. NSAIDs, may control the inflammation but do not influence the outcome of the disease.

2. Corticosteroids are indicated in acute systemic involvement, severe joint disease refractory to other drugs and uveitis. They do not alter the long-term outcome of the disease.

3. Disease-modifying agents, e.g. hydroxychloroquine, D-penicillamine, gold. Used in progressive polyarticular JCA.

4. Cytotoxic agents are rarely indicated. Uses include very severe, progressive disease and amyloidosis.

5. Physiotherapy and splinting to maintain function and prevent contractures. Splinting may relieve pain in an acutely inflamed joint.

6. Surgery, e.g. soft-tissue release.

Rheumatic fever

Rheumatic fever develops as an abnormal immunological reaction to a β-haemolytic streptococcal infection. The revised Jones criteria require two major features, or one major and two minor features plus evidence of streptococcal infection, to make the diagnosis.

Diagnosis

1. Major features. Carditis, migratory polyarthritis, chorea, erythema marginatum, subcutaneous nodules.

2. Minor features. Previous rheumatic fever, arthralgia, fever, raised ESR or CRP, leucocytosis, prolonged PR interval on ECG.

3. Evidence of streptococcal infection. Raised ASO titre, positive throat swab culture.

Carditis may present with a pericardial rub or effusion (pericarditis), arrhythymias or heart failure (myocarditis), or a heart murmur (endocarditis). Up to 50% will have have long-term cardiovascular disease (e.g. valvular damage).

Treatment

Treatment is with bed rest, anti-inflammatory drugs (NSAIDs) and penicillin. Rheumatic fever can recur and prophylactic penicillin should be given until well into adult life.

Further reading

Griffiths ID, Craft AW. Management of juvenile chronic arthritis. *Hospital Update,* 1988: 1372–84.

Related topics of interest

Limping and joint disorders (p. 205)
Purpura and bruising (p. 278)

POLYURIA AND RENAL TUBULAR DISORDERS

Polyuria is the passing of excessive amounts of urine. It is a frequent presenting symptom which may indicate renal, metabolic or endocrine disease.

Aetiology

(a) Excessive fluid intake

(b) Renal disease

- Renal tubular disorders.
- Acute tubular necrosis – recovery phase.
- Chronic renal failure.
- Nephrogenic diabetes insipidus.

(c) Metabolic/endocrine disorders

- Diabetes mellitus.
- Diabetes insipidus.
- Hypercalcaemia.
- Adrenogenital syndrome.
- Conn's syndrome.

(d) Drugs, e.g. antihistamines

Diabetes insipidus

Diabetes insipidus (DI) is the inability to concentrate urine because of total or partial vasopressin (antidiuretic hormone, ADH) deficiency. Large amounts of dilute urine are passed and polydipsia maintains the plasma osmolality in the normal range. If oral intake is inadequate, hypernatraemic dehydration results. ADH is synthesized in the supraoptic hypothalamic neurones and then stored in the posterior pituitary. Secretion is controlled by plasma osmolality. Central DI may result from hypothalamic disease (e.g. craniopharyngioma, histiocytosis X) but is often idiopathic. It may be associated with other pituitary hormone deficiencies. Treatment is with synthetic ADH (DDAVP) given by nasal spray. Unresponsiveness of the distal tubules and collecting ducts of the kidney to ADH may also cause polyuria and polydipsia (nephrogenic DI). It is present at birth and occurs mainly in boys. No specific treatment is available other than adequate fluid replacement and reduction of dietary solute load.

Investigation of polyuria

1. History and examination. Age, duration of symptoms, headaches and visual disturbance, rickets, family history, etc.

2. Urinalysis. Glucose, amino acids, pH to exclude diabetes mellitus and most renal tubular disorders.

3. Paired urine and plasma osmolalities and a water deprivation test to distinguish between central and nephrogenic DI and psychogenic polydipsia.

(*Note*: Water deprivation is potentially dangerous in DI and the test should be carefully monitored). Plasma osmolality will be normal or raised in both types of DI, and normal or low in psychogenic polydipsia (normal = 275–295 mosmol/l). The urine osmolality is inappropriately low (50–200 mosmol/l) in DI with failure to concentrate >300 mosmol/l following water deprivation (normally concentrate to >750 mosmol/l). DDAVP raises the urine osmolality in central DI but not in nephrogenic DI. In psychogenic polydipsia urine osmolality is appropriately low and, although concentrating ability may be impaired, it should still rise to >500 mosmol/l on water deprivation.

Renal tubular disorders

Eighty per cent of the glomerular filtrate is reabsorbed in the proximal tubules of the kidney (mainly sodium, chloride, phosphate, glucose, amino acids and water). Bicarbonate is regenerated by carbonic anhydrase and reabsorbed in exchange for hydrogen ions. Further sodium and water are reabsorbed in exchange for potassium or hydrogen ions in the distal tubules under the control of the renin–angiotensin system. Defects in this process of absorption and secretion may be confined to a specific transport system or cause a generalized transport defect. Many of these defects are genetically determined.

Problems

- Electrolyte imbalance.
- Polyuria, polydipsia.
- Failure to thrive.
- Muscle weakness.
- Rickets.

Aetiology

1. Proximal tubule defects

(a) Generalized transport defects

- Fanconi's syndrome
  - Primary
    - Cystinosis.
    - Adult-type.
  - Secondary
    - Heavy metals (e.g. lead, cadmium).
    - Wilson's disease.
    - Hereditary fructose intolerance.
    - Galactosaemia.
    - Tetracycline which has deteriorated.

Tyrosinosis.

- Lowe's (oculo-cerebro-renal) syndrome.

(b) Specific transport defects

- Cystinuria. This is a disorder of intestinal absorption and renal tubular reabsorption of dibasic amino acids (cystine, lysine, arginine, ornithine). There is a risk of multiple calculi formation. It occurs in 1 in 600 of the population, but calculi form in only 3%.
- Hartnup disease.
- Familial hypophosphataemic rickets.
- Renal tubular acidosis type 2 (proximal RTA).

2. Distal tubule defects

- Nephrogenic diabetes insipidus.
- Renal tubular acidosis type 1 (distal RTA).

Renal tubular acidosis (RTA)

Distal RTA

This is much more common than proximal RTA. It is due to failure of the distal tubules to excrete hydrogen ions. It is usually primary but may be secondary, e.g. following vitamin D intoxication, amphotericin toxicity or renal transplantation. Features include:

- Hyperchloraemic acidosis.
- Nephrocalcinosis.
- Rickets secondary to bone resorption to buffer the acidosis.
- Failure to thrive and weakness.
- Hyponatraemia and hypokalaemia due to increased urinary loss.
- Polyuria.
- Inappropriately alkaline urine and inability to acidify urine (ammonium chloride test).

Proximal RTA

This is due to increased bicarbonate loss secondary to a defect in bicarbonate reabsorption. It may be primary but often occurs as part of Fanconi's syndrome. It is usually sporadic but may be inherited. Primary proximal RTA with no associated glycosuria or aminoaciduria is more common in boys, and usually presents in infancy with failure to thrive, muscle weakness and persistent vomiting. There is a hyperchloraemic acidosis with appropriately acidic urine. Nephrocalcinosis is very rare.

Fanconi's syndrome

This is a generalized defect of proximal tubular function resulting in excess glucose, amino acids, uric acid, phosphate, sodium, potassium, bicarbonate and protein in the urine. The defect may be primary or secondary to tubular damage in other conditions (see above). Presenting features include failure to thrive, anorexia, vomiting, polyuria, polydipsia, rickets, muscle weakness, hypotonia, paralytic ileus and acute dehydration with metabolic acidosis. The primary defect is divided into two forms which affect both sexes equally.

Cystinosis (Lignac–Fanconi syndrome) This is a severe autosomal recessive condition in which cystine is deposited in many tissues. Renal tubular deposition and progressive glomerular damage lead to renal failure in early life. Cystine crystals can be identified in bone marrow, on renal biopsy or by slit-lamp examination of the cornea.

Adult-type Fanconi's syndrome The inheritance of this condition is less clear. It is not associated with cystinosis, and progression to renal failure is slow.

Further reading

Chesney RW. Clinical study of renal tubular disease. In: Barakat AY, ed. *Renal Disease in Children*. New York: Springer Verlag, 1989; 185–206.

Rodriguez-Soriano J. Acid–base disturbances. In: Barakat AY, ed. *Renal Disease in Children*. New York: Springer Verlag, 1989; 234–68.

Related topics of interest

Acute renal failure (p. 10)
Adrenal disorders (p. 14)
Chronic renal failure (p. 75)
Diabetes mellitus (p. 118)

PREMATURITY

Prematurity is defined as delivery before 37 completed weeks of gestation and is the second most common cause of neonatal death after malformation. Survival rates have improved with advances in perinatal and neonatal intensive care over the last 20 years. The majority of infants over 32 weeks' gestation who weigh over 1500 g are now expected to survive. Some infants delivered as early as 22 weeks have survived with intensive care, but those of less than 500 g or 22 weeks are considered unviable in most centres. Low birth weight (LBW < 2500 g) infants account for about 6% of the total births, and very low birth weight infants (VLBW < 1500 g) make up less than 1% of the total. An infant may be both small for gestational age and preterm. Premature delivery is due to either preterm labour or obstetric intervention when the risks to the mother or the fetus of continuing the pregnancy are very high, e.g. pre-eclamptic toxaemia (PET), severe IUGR. Prematurity is more common in young mothers, smokers, those with poor socioeconomic backgrounds and those with a previous history of a preterm delivery.

Problems

- Hypothermia.
- Hypoglycaemia.
- Respiratory distress.
- Nutrition.
- Jaundice.
- Susceptibility to infections.
- Necrotizing enterocolitis.
- Patent ductus arteriosus.
- Haemorrhagic and ischaemic cerebral lesions.
- Retinopathy of prematurity.
- Neurodevelopmental outcome.
- Parent–child bonding.

Aetiology of preterm labour

- Idiopathic (~ 50%).
- Multiple pregnancy.
- Antepartum haemorrhage.
- Premature rupture of membranes.
- Polyhydramnios.
- Cervical incompetence, e.g. previous surgery, congenital.
- Uterine malformations, e.g. bicornuate uterus.
- Maternal illness, e.g. diabetes mellitus, pre-eclampsia (PET), intercurrent viral infection, UTI.
- Infection, e.g. *Listeria.* Other organisms (e.g. *Mycoplasma hominis, Ureaplasma urealyticum*) may have a role in the aetiology of premature rupture of membranes.

Survival

For infants born after 32 weeks' gestation without congenital abnormality there is over 95% survival. Those of between 30 and 32 weeks have a survival rate of 90–95%, and this falls to 75–80% for those of between 28 and 30 weeks. Infants of 28 weeks' gestation or less show improved survival when treated in recognized neonatal intensive care units. At 26 weeks ~ 45% survive, at 24 weeks only ~ 30% survive and at 23 weeks less than 10% survive.

Intraventricular haemorrhage (IVH) and hypoxic–ischaemic encephalopathy (HIE)

Impaired autoregulation of cerebral blood flow, cardiopulmonary instability, sepsis and metabolic derangements predispose the preterm infant to cerebral haemorrhage and ischaemia. Haemorrhage arises from friable capillaries of the germinal layer and may be limited to this layer, or extend into the ventricular space or the surrounding parenchyma. In 25–30% of infants < 1500 g an IVH will develop, and 90% occur by the third day of life. Many are small, clinically silent and have no sequelae. A fall in the haematocrit may be the only clue. Large haemorrhages may result in death or neurological sequelae as a result of periventricular cerebral injury or progressive hydrocephalus. Thirty per cent of IVHs increase in size in the first week of life, and post-haemorrhagic hydrocephalus is common following moderate and large IVHs. Hydrocephalus becomes static or resolves spontaneously in >50% of cases but, if progressive, serial lumbar punctures or ventricular shunts may be needed. There is no evidence that Caesarean delivery reduces the incidence of IVH, and the role of prophylactic treatments (e.g. ethamsylate, vitamin E) is unclear.

HIE is associated with perinatal and post-natal asphyxia. Cerebral ischaemia may result from a period of hypotension and occurs particularly in arterial watershed areas (e.g. periventricular leucomalacia, PVL).

Retinopathy of prematurity (ROP)

ROP is a vasoproliferative retinopathy which can result in retinal detachment and blindness. The underlying mechanism by which damage to the immature retina occurs is still poorly understood. The association with hyperoxia has been well documented, but there is no absolute level of arterial oxygen tension above which ROP will occur, and it is occasionally seen in infants who have never received supplemental oxygen. Risk factors include low gestation and birth weight (particularly < 1000 g), and multiple birth. It is associated with ischaemic and haemorrhagic cerebral lesions and almost all affected infants have neurological impairment at follow-up. All infants under 34 weeks' gestation and < 1500 g should be screened with indirect ophthalmology at

7–9 weeks of age. Acute changes are seen in 70–80% of infants < 1000 g, but many regress. Progressive retinopathy is treated with cryotherapy. The results of surgery once the retina has become detached are poor.

Neurodevelopmental outcome

About 15–20% of surviving infants of < 1000 g and about 10% of those of 1000–1500 g have a moderately serious disability. Between 8 and 10% of surviving infants of 24–25 weeks' gestation have a major impairment (e.g. cerebral palsy, blindness). Prematurity is particularly associated with spastic diplegia and relative sparing of intellect. There is no conclusive evidence about the cause of neurological sequelae, but the sickest infants are at the greatest risk. Ischaemic and moderate to severe haemorrhagic cerebral lesions detected on US scan are predictive of a worse neurological outcome. Magnetic resonance studies have shown an association between deranged cerebral energy metabolism in the first week and neurodevelopmental impairment at 1 year of age. The effects of early diet on long-term neurodevelopment and intellect are controversial.

Further reading

Campbell AGM. Follow-up studies. In: Roberton NRC, ed. *Textbook of Neonatology,* 2nd edn. Edinburgh: Churchill Livingstone, 1992: 43–74.

Ventriculomegaly trial group. Randomised trial of early tapping in neonatal posthaemorrhagic ventricular dilatation. *Archives of Disease in Childhood*, 1990: **65:** 3–10.

Related topics of interest

Neonatal jaundice (p. 231)
Neonatal respiratory distress (p. 235)

PUBERTY – PRECOCIOUS AND DELAYED

Puberty is a period of changing patterns of hormone secretion resulting in physical and physiological changes which culminate in sexual maturation and fertility. There is great variability in the onset of these changes, but 95% of girls will have commenced puberty by the age of 13 years and 95% of boys will have commenced before the age of 14 years. Precocious puberty may be central (true) resulting from activation of the hypothalamic–pituitary–gonadal axis, or peripheral (false), involving an abnormal pathway and usually resulting in incomplete maturity. Precocious puberty is more common in girls (10:1) but is usually idiopathic. In boys, the majority will have a pathological cause and they therefore require more careful investigation. Delayed puberty is more common in boys and is usually constitutional.

Normal puberty

Normal puberty requires the secretion of both growth hormone and sex steroids. Pulsatile release of gonadotrophin-releasing hormone (GnRH) from the hypothalamus is initially nocturnal, and stimulates release of luteinizing hormone (LH) and follicle-stimulating hormone (FSH) from the pituitary. Spikes of hormone release are detected before clinical signs of puberty, and these increase in amplitude and frequency as puberty progresses. By late puberty, pulsatile LH and FSH secretion occurs throughout the 24 hours. The average duration of puberty is 2–3 years.

Girls

LH stimulates ovulation and FSH stimulates release of oestradiol from the ovaries. Breast development is the first sign of puberty. The growth spurt occurs early in puberty with the attainment of breast stage B2, and continues to menarche and breast stage B4.

Boys

LH stimulates release of testosterone from the testes. FSH stimulates spermatogenesis and thus determines testicular volume. Puberty starts approximately 0.6 years later than in girls, the first indication being increase of testicular volume to 4 ml. The growth spurt occurs later in puberty than in girls, when the testicular volume has reached 10–12 ml.

Children requiring investigation

- Girls commencing puberty < 8 years.
- Boys commencing puberty < 9 years.
- Lack of pubertal development in a girl > 14 years or a boy > 15 years.
- Puberty which pursues an abnormal sequence or stops before completion.

Precocious puberty

Aetiology

(a) Idiopathic.

(b) Central/true

- Intracranial tumours.
- Hydrocephalus.
- Radiotherapy (e.g. CNS prophylaxis in treatment of leukaemia).
- Post-infection, e.g. meningitis
- Neurofibromatosis.
- McCune–Albright syndrome (*café au lait* patches, polyostotic fibrous dysplasia, precocious puberty; commoner in girls; GnRH independent).
- Ectopic gonadotrophin release, e.g. hepatoblastoma.

(c) Peripheral/false

- Ovarian, testicular or adrenal tumours.
- Virilizing congenital adrenal hyperplasia.
- Exogenous sex steroids.

Clinical features

If a child shows signs of early pubertal development it is important to decide if the normal sequence of physical development (consonant with puberty) has occurred. This suggests stimulation of the hypothalamic–pituitary–gonadal axis (true precocious puberty) usually by premature activation of GnRH pulsatility. In the majority of girls this is idiopathic, but in boys a central precipitating cause should be suspected. If there is lack of consonance (e.g. pubic hair and acne without breast development) peripheral causes should be considered (false precocious puberty). Signs of virilization associated with hypertension suggest an adrenal cause. Early breast development (premature thelarche) with no other signs of puberty (i.e. normal growth rate, no pubic hair, appropriate bone age) is a benign condition due to isolated pulsatile FSH secretion which requires no treatment. Gynaecomastia in boys of pubertal age is extremely common and usually subsides as puberty progresses. Isolated vaginal bleeding may be due to a foreign body.

Investigation

- Growth assessment and pubertal staging.
- LH, FSH profiles. Pulsatile release is seen in true precocious puberty, low levels in false precocious puberty.
- GnRH stimulation test. Elevated basal FSH and LH levels with exaggerated response to LHRH in true precocious puberty.

- Pelvic US. Multicystic appearance of pubertal ovaries.
- Bone age. Left wrist.
- CNS imaging. Always in boys, only if neurological signs in girls.
- Plasma and urinary androgens, if virilized. Plasma oedradiol, prolactin if oestrogenized.
- Thyroid function. Thyroid-releasing hormone stimulates prolactin and FSH release as well as thyroid-stimulating hormone (TSH), so breast development in girls or testicular enlargement in boys may be seen in hypothyroidism.

Management

If a cause is identified it may be removed, although the endocrine abnormality may not resolve. GnRH analogues are used to arrest puberty by blocking pulsatile release of FSH and LH from the pituitary. They can be given intranasally, subcutaneously or by depot injection. Growth hormone may be given to accelerate growth in an attempt to compensate for advanced skeletal maturity.

Delayed puberty

Aetiology

- Constitutional in the majority.
- Primary gonadal failure (e.g. Turner's syndrome (XO), irradiation, chemotherapy).
- Secondary gonadal failure (e.g. hypogonadotrophic hypogonadism, pituitary or hypothalamic tumours).
- Chronic illness, anorexia nervosa.

Clinical features

If there are no signs of puberty, general history and examination may identify the presence of chronic illness. Constitutional pubertal delay is usually associated with short stature, whereas normal stature is usual in hypogonadotrophic hypogonadism. Features of Turner's syndrome may be present (see Chromosomal abnormalities, p. 68). Incomplete or abnormal puberty (lack of consonance) is uncommon and causes include disorders of steroidogenesis and ACTH deficiency.

Investigation

- LH, FSH. Low in hypothalamo-pituitary problems (secondary gonadal failure due to decreased gonadotrophin release) and also in constitutional delay. Elevated in primary gonadal failure, e.g. Turner's syndrome. The GnRH stimulation test is unreliable in distinguishing hypogonadotrophic gonadism from simple pubertal delay.

- Bone age.
- Thyroid function tests.
- Chromosomes – particularly in girls.

Management

If constitutional delay is causing psychological distress, puberty may be induced with sex steroids in low doses. Hormone replacement treatment is indicated in Turner's syndrome.

Further reading

Brook CGD. Puberty. In: *A Guide to the Practice of Paediatric Endocrinology.* Cambridge University Press, 1993; 55–82.

Related topics of interest

Chromosomal abnormalities (p. 68)
Growth – short and tall stature (p. 149)

PURPURA AND BRUISING

The commonest causes of purpura and bruising in childhood are trauma, Henoch–Schönlein purpura (HSP), idiopathic thrombocytopenic purpura (ITP) and leukaemia. Less common but important causes are meningococcal septicaemia and inherited coagulopathies, which are discussed elsewhere. The aetiology may be thrombocytopenia, a vascular problem, or a coagulation factor deficiency. Conjunctival haemorrhages and facial petechiae are common following raised intrathoracic pressure (e.g. whooping cough, vomiting, newborn following vaginal delivery) but may occur spontaneously in leukaemia and other bleeding disorders.

Aetiology

(a) Vascular

- Trauma (accidental, non-accidental).
- HSP.
- Infection, e.g. *Meningococcus*.
- Rare, e.g. hereditary haemorrhagic telangiectasia.

(b) Thrombocytopenia ($< 150 \times 10^9/l$)

- ITP.
- Hypersplenism, e.g. haematological (spherocytosis), malignancy (lymphoma), infections (infectious mononucleosis), metabolic (Gaucher's disease).
- Consumption, e.g. DIC, giant haemangioma.
- Decreased production, e.g. bone marrow infiltration (leukaemia, neuroblastoma), drugs (chloramphenicol).
- Immune, e.g. drugs (quinine), autoimmune (SLE).

(c) Platelet dysfunction

- Drugs, e.g. aspirin.
- Metabolic, e.g. uraemia.
- Inherited, e.g. von Willebrand's, thrombasthenia.

(d) Coagulopathy

- Inherited (haemophilia, Christmas disease, von Willebrand's disease).
- Acquired (DIC, haemorrhagic disease of the newborn, liver disease).

Investigation

- *FBC*. The two most likely causes of thrombocytopenia are ITP and leukaemia. The platelet count is usually $< 30 \times 10^9/l$ before abnormal bruising and petechiae occur.
- *Partial thromboplastin test (PTT)*. This is a test of the intrinsic clotting factors (VIII, IX, XI, XII). It is abnormal in haemophilia, Christmas disease and von Willebrand's disease.
- *Prothrombin time (PT)*. This is a test of the extrinsic clotting factors (II, V, VII, X). It is normal in haemophilia and Christmas disease, prolonged in liver disease and vitamin K deficiency.

- *Bleeding time*. A prolonged bleeding time in the presence of a normal platelet count suggests platelet dysfunction, e.g. von Willebrand's disease, thrombasthenia.

Henoch–Schönlein purpura (HSP)

Problems

- Rash.
- Joint pains.
- Glomerulonephritis.
- Abdominal pain.

Clinical features

HSP is a diffuse, self-limiting allergic vasculitis of unknown aetiology (also known as anaphylactoid purpura). It occurs particularly in pre-school children, and is twice as common in boys. Peak incidence is in the winter. The first symptom in most (>50%) is the appearance of a purpuric rash in a relatively well child. The characteristic distribution is over the extensor surfaces of the limbs and buttocks, but it may involve the trunk and occasionally the face. It begins as a raised urticarial rash, rapidly progressing to purpura. Pain and swelling of one or more joints (particularly ankles and knees) occurs in two-thirds of cases and is the presenting symptom in 25%. Gastrointestinal involvement is common with colicky abdominal pain due to haemorrhage into the gut wall. Less common problems include melaena, bloody diarrhoea and intussusception. Renal involvement is common, with microscopic haematuria occurring in >70%. Features of acute glomerulonephritis (macroscopic haematuria, oliguria, hypertension) or nephrotic syndrome (oedema, proteinuria) develop in ~ 25%, and rare cases progress to end-stage renal failure. CNS symptoms (fits, behaviour disturbance) are uncommon.

Investigation

There are no specific laboratory findings and diagnosis is made on the clinical presentation. The platelet count and clotting studies are normal.

Management

Treatment is symptomatic with analgesics for joint and abdominal pain, plus bed rest if joint pain is marked. A short course of steroids can be used for severe abdominal pain. Urinalysis for blood and protein, and regular blood pressure monitoring are important. The benefits of specific therapies for severe renal involvement (aspirin, pulsed methylprednisolone, plasma exchange, cyclophosphamide) are unclear.

Outcome

HSP runs a variable course with the majority settling within 2–6 weeks. Relapses can occur but rarely beyond 1 year. Microscopic haematuria may persist for several months or even years. The renal function of those with acute nephritis or nephrosis may continue to deteriorate even after 5 years so long-term surveillance is essential.

Idiopathic (immune) thrombocytopenic purpura (ITP)

Problems

- Chronic blood loss.
- Intracranial haemorrhage (risk < 1%).
- Side-effects of treatment.

Clinical features

ITP presents with acute onset of bruising, purpura and petechiae. Less commonly mucosal bleeding occurs. Eighty per cent report a preceding viral illness, but the child is usually well at presentation and associated symptoms, e.g. pallor, lethargy, bone pain, should suggest leukaemia rather than ITP. In 75% the platelet count returns to normal within 3 months, but in 10–20% thrombocytopenia persists beyond 6 months (chronic ITP). Mortality is low and is usually due to intracranial haemorrhage.

Investigation

- FBC. Thrombocytopenia often $< 10 \times 10^9/l$.
- Immunoglobulins. IgG, IgA. Platelet-associated IgG is almost always present but difficult to detect with very low platelet counts. Platelet antibody tests are usually unhelpful.
- Viral titres, e.g. Epstein–Barr virus, rubella, CMV (if under 1 year).
- Bone marrow aspiration is probably not necessary if the platelet count is $> 30 \times 10^9/l$, but the child should be observed and bone marrow aspiration performed if spontaneous remission does not occur within 2–3 weeks. Bone marrow aspiration is essential if any features suggestive of marrow infiltration or aplasia are present, or if treatment (particularly steroids) is planned. In ITP there are normal or increased numbers of megakaryocytes.
- Anti-nuclear factor, anti-DNA antibodies. Exclude SLE if ITP becomes chronic.

Management

Children are usually admitted to hospital to make the diagnosis. There is no need to keep them in hospital or restrict activities other than to try to prevent fights and vigorous knockabouts, particularly when the platelet count is

< 10×10^9/l. No treatment is necessary if there is no mucosal bleeding or serious haemorrhage, but a short course of low-dose steroids may lead to slightly faster recovery of the platelet count. Bone marrow aspiration to exclude leukaemia must be performed before starting treatment. Chronic ITP has been treated with pulses of high-dose methylprednisolone, immunoglobulin infusions and splenectomy, but in view of the side-effects of these treatments they are best reserved for serious haemorrhagic emergencies, together with massive allogeneic platelet transfusion. ITP may remit at any time even after years, and serious haemorrhage is rare. Treatment is symptomatic, not curative, so the child should be treated on the basis of symptoms and not the platelet count.

Further reading

Eden OB, Lilleyman JS on behalf of the British Paediatric Haematology Group. Guidelines for management of idiopathic thrombocytopenic purpura. *Archives of Disease in Childhood,* 1992; **67:** 1056–8.

Related topics of interest

Child abuse (p. 59)
Glomerulonephritis (p. 146)
Haemolytic uraemic syndrome (p. 156)
Haemophilia and Christmas disease (p. 159)
Leukaemia and lymphoma (p. 201)

PYREXIA OF UNKNOWN ORIGIN

Pyrexia of unknown origin (PUO) may be defined as a fever of over one week's duration which remains unexplained after at least one week of hospital investigation. The term is often used more loosely in the paediatric age group if the cause of a fever is not immediately apparent. Fever is a normal part of the body's response to infection, although dehydration and febrile convulsions are complications of the fever itself.

Aetiology

- Infection.
- Collagen vascular disease.
- Malignancy.
- Miscellaneous.

1. Infection. Bacterial infection may cause systemic infection or may be localized, e.g. abscess. Pyogenic liver abscess, subphrenic or perinephric collections may be difficult to diagnose. Tuberculosis, particularly non-pulmonary, may present with PUO. Viral causes, e.g Epstein–Barr virus, CMV, should be considered. Chlamydial, rickettsial, spirochaete, fungal and parasitic infections may be responsible. Infective endocarditis must be considered in a child with a PUO and a heart defect, although it can occur in structurally normal hearts.

2. Collagen vascular disease. A remittent swinging fever may precede the joint manifestations of Still's disease, and a macular rash may be seen at the height of the fever. Fever is also a feature of systemic lupus erythematosus, when it is accompanied by malaise, weight loss, arthralgia or arthritis and the characteristic 'butterfly' facial rash. Rarely, childhood polyarteritis nodosa may be responsible.

3. Malignancy. Leukaemia, lymphoma, neuroblastoma and some solid tumours may be associated with fever at presentation, although it is unusual for fever to be the only presenting symptom.

4. Miscellaneous. Inflammatory bowel disease, thyrotoxicosis and diabetes insipidus may rarely present with a fever. Kawasaki disease features a fever of more than 5 days' duration, accompanied by other defined clinical features, the most sinister of which is coronary artery aneurysms. Riley–Day syndrome (familial dysautonomia), anhidrotic ectodermal dysplasia and familial Mediterranean fever are other rare causes. The possibilities of drug-induced pyrexia or Munchausen syndrome by proxy must also be considered.

History

A careful history must be taken of the fever and associated symptoms, e.g. cough, rash, night sweats, abdominal or joint pain, etc. There may be a recent history of travel, insect bites, contact with animals or ingestion of untreated milk (tuberculosis or brucellosis). Previous drug administration may be the cause of the fever, or it may be modifying an infective process. Family history may provide diagnostic information in the rarer inherited conditions such as Riley–Day syndrome.

Examination

The pattern of fever may be helpful, e.g. a sustained high fever in bacteraemia or a relapsing fever in infective endocarditis. Meticulous examination should be performed to look for diagnostic clues and should be repeated at frequent intervals to look for fresh signs, e.g. Roth spots in infective endocarditis. Ears, nose and throat should be closely examined, and the presence of rashes, lymphadenopathy or localized facial tenderness noted. The eyes may show petechial haemorrhages in meningococcaemia, or conjunctivitis in tuberculosis or connective tissue disease. Absent tear production and corneal reflex occurs in the Riley–Day syndrome. There may be changing murmurs in infective endocarditis, and nailbeds may reveal splinter haemorrhages. Chest, abdominal and rectal examination may provide localizing signs. There may be localized bony points of tenderness in osteomyelitis, and generalized tenderness of muscles in viral, bacterial or parasitic myositis.

Investigation

- FBC, including a blood film. Thick and thin films may be needed if malaria is suspected.
- ESR. Raised in active inflammation.
- Bone marrow. Exclude infiltration by leukaemia, lymphoma, tuberculosis, metastatic malignancy.
- Blood cultures. Repeat frequently, especially if infective endocarditis is suspected.
- Bacterial antigen detection, if previous antibiotics given.
- Urine, stool and sputum culture, gastric aspirates or early morning urines for acid–alcohol-fast bacilli.
- Mantoux test.
- Serum for viral serology.
- Imaging. Chest radiography, abdominal ultrasound, CT or MRI imaging may show occult sites of infection, and

may guide the diagnostic aspiration of localized pus. Gallium- or technetium-labelled white cell scans may help in localizing infection. Contrast barium studies may be needed if inflammatory bowel disease is suspected.

- Liver or lymph node biopsy is occasionally required.

Management

In general, antibiotic treatment should be withheld until the diagnosis has been established, but this is clearly inappropriate in sick or deteriorating children, and supportive management combined with best guess antimicrobial therapy is then required. Dehydration should be avoided. Antipyretic agents commonly used are paracetamol and, more recently, ibuprofen.

Further reading

Hull D. Fever – the fire of life. *Archives of Disease in Childhood,* 1989; **64:** 1741–7.

Related topics of interest

Childhood exanthemata (p. 62)
Leukaemia and lymphoma (p. 201)
Lymphadenopathy (p. 214)
Malignancy in childhood (p. 217)

RASHES AND NAEVI

Problems

- Cosmetic.
- Pruritus.
- Infection.
- Malignancy.

Haemangiomata

The 'stork mark' is a fine, non-raised capillary naevus commonly found in the nape of the neck, or in the midline over eyelids, forehead or upper nose. No treatment is required – fading is usual.

The port-wine stain is a darker capillary naevus which does not fade and may be cosmetically disfiguring. If the area of the ophthalmic division of the trigeminal nerve is affected, there may be an associated intracranial haemangioma (Sturge–Weber syndrome). Laser treatment may now be used.

Cavernous haemangioma may be superficial ('strawberry marks') or deep, often growing rapidly during the first year of life, with complications of bleeding or infection. Spontaneous regression occurs within 4–5 years, although large lesions may lead to platelet sequestration and need steroid therapy. γ-Interferon has also recently been used.

Spider naevi consist of vessels radiating from a central arteriole and are commonly found on the upper body and face. Multiple spider naevi are associated with chronic liver disease.

Pigmented naevi

Café au lait patches are hyperpigmented macular lesions and are usually innocent. Six or more lesions of diameter greater than 1.5 cm are found in association with neurofibromatosis.

Mongolian patches are flat lesions of grey or blue/black, found on the buttocks and lumbosacral spine of Negro or Oriental babies, and occasionally in Caucasian races. Lesions may be mistaken for bruising. They fade by the end of the first decade.

Melanocytic naevi (moles) are very commonly found on the face, back and neck. There is a small risk of malignancy in later life. The giant pigmented naevus, often of 'bathing trunk' distribution over lower abdomen and buttocks, is at particular risk of malignant change.

Epithelial naevi

Verrucous naevi may be localized, linear and solitary with a brown, warty surface. Extensive areas of verrucous naevus may be associated with developmental defects of other systems – the epidermal naevus syndrome.

Sebaceous naevi occur over the scalp as smooth, raised waxy plaques. Basal cell carcinoma may occur in adult life, therefore excision is advised.

Napkin rashes

Napkin dermatitis is due to prolonged contact with wet napkins, producing erythema and ulcers on the exposed parts, sparing the flexures. Frequent changing, careful hygiene, barrier creams and exposure are usually effective measures.

Napkin candidiasis often complicates napkin dermatitis or oral antibiotic therapy, presenting as an erythematous rash with characteristic satellite lesions. Treatment is with nystatin preparations – steroid combinations are effective if there is inflammation.

Infantile seborrhoeic dermatitis may occur in the first 2 months of life as an erythematous and scaly eruption in the groin, scalp (cradle cap) and neck. Skinfolds are affected. Treatment with a mild steroid cream and antiseptic is generally effective. Langerhans' histiocytosis may present with a similar rash, although usually in an older infant.

Psoriasis

Small papular lesions with silvery scales most commonly occur in childhood after a streptococcal sore throat (guttate psoriasis), and may clear spontaneously over 2–3 months, although recurrences are likely. Knees, elbows and scalp are commonly affected. The Köebner phenomenon (lesions along the site of skin injury) may be seen. Chronic psoriasis is less common in childhood, producing plaques and rarely nail changes. Treatment includes bland ointments, coal tar preparations, dithranol and UV light. Methotrexate may be needed in very severe cases.

Viral eruptions

Molluscum contagiosum is due to a pox virus, causing clusters of pearly papules with central umbilication. Spontaneous resolution is the rule, although cautery or freezing may be used.

Warts, caused by human papillomavirus, usually affect the fingers and soles. They may be painful if they are on the soles of the feet or if there is secondary infection. Genital warts occasionally suggest child sexual abuse. Warts may be treated with liquid nitrogen or podophyllin paint if necessary.

Pityriasis rosea is probably a viral eruption. A herald patch precedes the rash by a few days. Itchy, scaly patches occur over the trunk, sometimes over the line of the ribs posteriorly. Mild topical steroid treatment for irritation may be needed.

Infestation

Scabies is caused by *Sarcoptes scabiei* and produces itching and an erythematous papular rash. Close family contact is usually responsible for spread. Burrows with scratch marks are found on wrists, interdigital spaces, face and genitalia. Treatment must be for the whole family with γ-benzene hexachloride. Bedclothes should be thoroughly washed.

Fungal infection

Candida

See Napkin rashes (p. 286).

Pityriasis versicolor

Pityriasis versicolor appears as light-brown scaly patches and may itch. The cause is a fungus, *Malasszia furfur*, and topical antifungal treatment should be effective.

Urticaria

Urticaria (nettle rash, hives) is an itchy erythematous rash, with weals, due to increased permeability of local capillaries.

There may be a history of other allergic phenomena. Physical agents (sunlight, mechanical pressure – dermographism), foods and drugs are precipitants, and management is by allergen avoidance and antihistamines.

Further reading

Verbov J, Morley N. *Colour Atlas of Paediatric Dermatology*. Lancaster: MTP Press, 1983.

Williams REA, Morley WN. Diagnosis of systemic disorders from skin signs. *Current Paediatrics*. 1992; **2:** 90–2.

Related topics of interest

Blistering conditions (p. 41)
Eczema (p. 127)
Neurocutaneous syndromes (p. 246)

SEIZURES

Seizures, faints and funny turns during childhood may cause diagnostic difficulty. An accurate diagnosis must be based on a reliable eye-witness account of the event, clinical examination, and selected investigations. Convulsions are seizures which are epileptic in origin, i.e. resulting from an abnormal and excessive discharge of activity of the cerebral neurones. Many 'attacks' in children are of non-epileptic origin, e.g. syncope, hyperventilation, hypoglycaemia, breath-holding attacks (white, reflex anoxic; or blue, cyanotic) or cardiac events (arrhythmias or outflow obstruction).

Problems

- Recurrent seizures.
- Status epilepticus.
- Side-effects of medication.
- Social and educational effects.

Clinical assessment

An eye-witness account of the convulsion is essential, to give details of apparent precipitating events (fever, flickering lights, trauma) and a description of the physical events, e.g. associated facial movements, urinary incontinence, tonic–clonic movements, etc. The child's age and birth, family and developmental history may give important diagnostic clues. If the convulsion is still in progress, priority must be given to airway management, preventing hypoxia and hypoglycaemia and cessation of the episode by drug therapy (see below). Further examination may reveal important signs, e.g. fever, stigmata of neurocutaneous syndromes, etc.

Investigation

According to the clinical picture, selected investigations may be necessary:

- FBC, e.g if infection suspected. The white cell count is often raised after a convulsion.
- U&E, calcium, glucose.
- EEG.
- CT or MRI imaging of brain.

Febrile convulsions

These are convulsions associated with non-cerebral infection in otherwise normal children, and occur between the ages of 6 months and 6 years in 3–5% of children. The convulsion may be simple, i.e. a generalized convulsion of less than 15 minutes' duration, occurring only once during a febrile illness in the absence of focal signs, or may be complex, i.e. a focal or prolonged convulsion, often repeated within the same illness, or leaving residual focal signs (Todd's paresis). Recurrence of febrile convulsions occurs in 30%. The risk of later epilepsy is increased if there are pre-existing neurodevelopmental abnormalities, if complex febrile convulsions occur, or if there is a family history of true epilepsy. Prophylactic anticonvulsants should therefore be considered for this group of children. The aims of management are to diagnose and treat the cause of fever (meningitis or UTI may be the source), assess risk factors for recurrence or further surveillance and to counsel the parents on first-aid management of fever and convulsions.

Infantile spasms (West's syndrome)

Repetitive flexor and/or extensor spasms ('salaam spasms') commence between 4 and 7 months, accompanied by developmental problems and a typically chaotic EEG pattern – hypsarrhythmia. In many cases there is a demonstrable brain abnormality, e.g. tuberous sclerosis. Early treatment is indicated, although attacks may be difficult to control. ACTH, steroids, benzodiazepines and, recently, vigabatrin have all been used.

Epilepsy

Epileptic seizures may be generalized from the onset, or partial (with or without secondary generalization).

Generalized seizures

1. Childhood absence epilepsy (petit mal). 'Absence attacks' occur for less than 10 seconds over 10 times daily, and may be induced by hyperventilation. Deteriorating school performance may be an early feature. Females are more commonly affected than males, often aged 6–8 years. The EEG shows typical 3/second spike-and-wave discharges. Response to sodium valproate or ethosuximide is usually good.

2. Grand mal – primary generalized epilepsy. Generalized tonic–clonic seizures may present between 5 and 10 years of age. There may be photosensitivity. The EEG shows bursts of spike waves between seizures. The remission response, normally with a single anticonvulsant drug (e.g. sodium valproate or carbamazepine), is 80–90%.

3. Myoclonic epilepsy. Sudden involuntary contraction occurs in a muscle or group of muscles, often with retained consciousness level. Various clinical pictures occur, e.g. atonic–akinetic 'drop attacks', symmetrical repetitive jerking, etc. However, those with Lennox–Gastaut syndrome suffer from an early onset (2–6 years) of intractable seizures with slow spike and wave activity on the EEG.

Partial seizures

These may be simple (consciousness retained) or complex (consciousness lost). Expression of seizures may be with motor, autonomic, psychic or somatosensory symptoms. Benign partial epilepsy, with centrotemporal (Rolandic) spikes on the EEG, accounts for 15% of childhood seizures, presenting during the first decade and remitting during adolescence. Complex partial seizures arising from the temporal or frontal areas account for 25% of paediatric seizures, and carbamazepine is usually effective (phenytoin, acetazolamide and occasionally surgery in resistant cases).

General principles of management

A firm diagnosis of the type of epilepsy should be made, as drug treatment may not be indicated, and prophylaxis after a single convulsion is rarely indicated. However, a recurrence of an attack within a month is an indication for prophylaxis. Single drug therapy should be used whenever possible (achieves a remission in 75% of cases), aiming to withdraw treatment as soon as possible after remission of seizures (usually 1–2 years).

Status epilepticus is a medical emergency, and priorities of management are to control the airway, ensure oxygenation and normoglycaemia while using drug treatment to stop the convulsion. Benzodiazepines, paraldehyde, phenytoin, phenobarbitone and chlormethiazole are drugs which may be used, but paralysis with mechanical ventilation for resistant status may be needed.

Further reading

Brown JK, Hussain IHMI. Status epilepticus. 2. Treatment. *Developmental Medicine and Child Neurology*, 1991; **33:** 97–109.

O'Donohoe NV. Use of antiepileptic drugs in childhood epilepsy. *Archives of Disease in Childhood,* 1991; **66:** 1173–5.

Wallace SJ. Convulsions and funny turns. *Current Paediatrics,* 1992; **2(2):** 63–7.

Related topics of interest

Developmental delay and regression (p. 115)

Neurocutaneous syndromes (p. 246)

SHOCK

Shock is inadequate tissue perfusion due to impaired cardiovascular function, which may be the end result of many disease processes. Impaired delivery of nutrients and oxygen to the tissues leads to anaerobic metabolism, acidosis, impaired cellular function and subsequent organ failure. Prompt recognition and resuscitation is vital to reverse this process and reduce both mortality and morbidity.

Problems

- Hypotension.
- Metabolic acidosis.
- Multisystem failure – hypoventilation, acute tubular necrosis, cerebral anoxia.

Aetiology

(a) Hypovolaemic shock

- Fluid and electrolyte loss
  - External, e.g. gastroenteritis, burns.
  - Internal, e.g. nephrotic syndrome.
- Blood loss, e.g. trauma, gastrointestinal bleed

(b) Cardiogenic shock

- Severe heart failure.

(c) Septicaemic shock

- Gram-negative sepsis, e.g. meningococcal septicaemia.
- Toxic shock syndrome.

(d) Drugs

- Overdose, e.g. barbiturates.
- Anaphylaxis.

Pathogenesis of septic shock

The clinical features of Gram-negative shock result from disruption of the normal homeostatic mechanisms of the vascular endothelium following activation of the host inflammatory responses by bacterial endotoxin. The normal endothelium maintains vascular permeability and inhibits thrombosis. Release of inflammatory mediators, including neutrophil-derived enzymes and cytokines (e.g. interleukin 2 and tumour necrosis factor) leads to three major processes:

- Increased vascular permeability resulting in hypovolaemia and oedema.
- Vasoconstriction of some vascular beds and dilatation of others leading to impaired organ function.
- Intravascular thrombosis with platelet and clotting factor consumption (DIC) resulting in further tissue hypoxia.

Prostacyclin (PGI_2) is an inhibitor of platelet aggregation and a potent vasodilator which is actively produced by the endothelium together with other important inhibitors of thrombosis such as nitric oxide. It has been suggested that a deficiency of PGI_2 may have a role in the pathogenesis of Gram-negative shock, but the role of exogenous PGI_2 as a therapeutic intervention has yet to be proven.

Clinical features

Shock may occur at any age but is most commonly seen in children under 5 years. The child lies quietly and is poorly responsive, or can be restless and confused. The peripheries are grey, mottled, cold and clammy with sluggish capillary filling. There may be central cyanosis and the breathing is often laboured with grunting. The pulse is thready, rapid and is difficult to feel peripherally. Hypotension is not a feature of early shock due to compensatory vasoconstriction but will eventually occur. Oliguria is usual. There may be decerebrate posturing if cerebral anoxia is marked or if septicaemia is associated with meningitis. The characteristic rash of meningococcal septicaemia should be readily recognized.

Toxic shock syndrome is an acute illness characterized by fever, mucous membrane hyperaemia, oedema, desquamating erythroderma and rapidly progressive hypotension with multisystem failure. It was originally associated with *Staphylococcus aureus* and became more widely recognized as an illness of young women who used tampons. However, a significant proportion of cases are not associated with menstruation, but with focal staphlococcal infection, and many of these cases occur in children. A toxic shock-like syndrome can also be caused by group A haemolytic streptococcus.

Investigation

- *FBC*. Hb is usually normal in acute blood loss. The packed cell volume (PCV) will be raised if the shock follows fluid loss. The WBC is often raised. Thrombocytopenia occurs with DIC.
- *Blood glucose*. Often low.
- *U&E, creatinine*. The degree of renal failure and electrolyte imbalance depends on the aetiology. Hyperkalaemia will be aggravated by acidosis.
- *Acid–base status*. Metabolic acidosis due to tissue hypoxia.
- *Blood cultures.*
- *Liver function tests*. Hepatic ischaemia leads to a rise in transaminases and ammonia.
- *Clotting*. DIC is invariably present in severe shock with consumption of clotting factors and elevation of FDPs. It is also associated with septicaemia, burns and major trauma.

- *Calcium, phosphate, magnesium.* Commonly deranged.
- *Chest radiography.* Cardiomegaly suggests cardiogenic shock. Pneumonia may be the primary illness or an aspiration pneumonia secondary to impaired consciousness level. In late shock pulmonary oedema and shock lung may develop.
- *Urine culture.* UTI may cause Gram-negative septicaemia and shock, particularly in infants.
- *Lumbar puncture.* Resuscitate and stabilize first.
- *Group and save serum.*

Management

1. Airway. Clear and maintain.

2. Breathing. Give oxygen, and if the patient is hypoventilating intubate and ventilate.

3. Circulation. Insert an i.v. line (peripheral, central, intraosseous), take bloods for investigations (including BM Stix) and give 20 ml/kg plasma bolus over 5–10 minutes. This may need to be repeated. Commence i.v. infusion – the type of fluid will depend on the underlying cause of shock.

4. Obtain a brief history of the illness from an accompanying adult (parent, paramedic), including any relevant past medical history, drugs, etc.

5. Examine the child noting level of consciousness (AVPU system), evidence of sepsis (rash, neck stiffness), dehydration (reduced skin turgor, sunken fontanelle), trauma, etc.

6. Arterial blood gas. Give i.v. sodium bicarbonate to reduce the degree of metabolic acidosis.

7. Admit to the intensive care area once stable.

8. Obtain a full history from the parents.

9. Further management will depend on the underlying cause of the circulatory failure. General measures to improve myocardial contractility include volume replacement and correction of acidosis, hypoxia, anaemia and electrolyte imbalance. In hypovolaemic shock the child will rapidly improve with plasma expansion, whereas in cardiogenic shock inotropes may be needed to improve myocardial function. If sepsis is suspected commence broad-spectrum antibiotics. The roles of prostacyclin and monoclonal antibodies to endotoxin in the treatment of septicaemic shock remain unproven.

Further reading

Levin M. Shock. In: Black D, ed. *Paediatric Emergencies,* 2nd edn. London: Butterworths, 1987: 87–116.

Lloyd-Thomas AR. Paediatric trauma. In: *ABC of Major Trauma*. London: BMJ Publications, 1991; 73–83.

Shock. In: Advanced Life Support Group. *Advanced Paediatric Life Support – the Practical Approach.* London: BMJ Publications, 1993: 73–86.

Related topics of interest

Acute renal failure (p. 10)
Burns (p. 44)
Cardiopulmonary resuscitation (p. 52)
Heart failure (p. 168)
Meningitis and encephalitis (p. 220)

SICKLE CELL ANAEMIA AND THALASSAEMIA SYNDROMES

Each molecule of haemoglobin normally contains four globin chains. At birth, 50–80% of haemoglobin is fetal haemoglobin (HbF = $\alpha_2\gamma_2$) and 15–40% is adult haemoglobin (HbA = $\alpha_2\beta_2$). A small proportion is $HbA_2 = \alpha_2\delta_2$. By 6 months of age HbF represents less than 5% of the total. By 3 years HbA represents 98–99% with one α-chain and one β-chain inherited from each parent. Sickle cell anaemia (HbSS) is a disorder of haemoglobin synthesis due to a single amino acid substitution (valine for glutamine) in the β-globin chain. Other β-chain variants include haemoglobin C (HbC). The thalassaemia syndromes result from impaired globin chain synthesis which may affect either the α- or β-chains. Each haemoglobinopathy can exist in the heterozygous or homozygous state. Mixed haemoglobinopathies, e.g. HbS/HbC or HbS/β-thalassaemia, occur and tend to be milder than HbSS.

Sickle cell anaemia

Problems

- Microcytic, hypochromic anaemia.
- Crises (painful, anaemic).
- Infections (*Pneumococcus, Salmonella*).
- Impaired renal concentrating ability.
- Family and social disruption.

Incidence

Sickle cell anaemia occurs mainly in those of Afro-Caribbean origin. In the UK, 15% of Negroes carry the gene, and it is a common inner city problem, with approximately 5000 people who are homozygous (sickle cell anaemia) and many more who are heterozygous (sickle cell trait).

Clinical features

High concentrations of HbS within red cells leads to aggregation of the HbS molecules when deoxygenated, resulting in distortion of the red cells (sickling), with subsequent increased blood viscosity and capillary obstruction. Dehydration, infection, acidosis and hypoxia exacerbate sickling.

Sickle cell anaemia usually presents after the age of 6 months (when levels of HbF have fallen) with painful swelling of the hands or feet (dactylitis), anaemia, icterus and hepatosplenomegaly. There is an increased risk of infections with encapsulated organisms because of splenic dysfunction (pneumococcal septicaemia, salmonella osteomyelitis), and these are a significant cause of mortality. Repeated splenic infarction results in less prominent splenomegaly in older children. Life is punctuated by painful and anaemic crises. Painful crises are due to vaso-occlusion, causing severe pain in any part of the body (particularly bones, muscle, abdomen). Severe abdominal pain may be mistaken for an 'acute abdomen'. Anaemic crises result from bone marrow aplasia (parvovirus), megaloblastic crisis (folate deficiency), visceral sequestration of red cells (e.g. hepatic, splenic) or haemolysis. Stroke occurs in ~ 5% of patients. Other complications include enuresis (impaired renal concentrating ability following repeated renal infarcts), leg ulcers, priapism, chest syndrome (chest pain, fever, leucocytosis, thrombocytopenia, high mortality) and aseptic necrosis of the hip. Delayed puberty is common. Heterozygotes rarely have problems but it is important to keep them well oxygenated during general anaesthesia.

Investigation

- *FBC.* Chronic haemolytic anaemia (microcytic, hypochromic). Hb is usually 6–9 g/dl, reticulocyte count raised. Sickle cells, target cells, and Howell–Jolly bodies may be seen on film. Blood can be deoxygenated as a screening test for sickling.
- *Haemoglobin electrophoresis*
  Homozygote: HbS 90–95%, HbF 5–10%.
  Heterozygote: HbS < 40%, HbA > 60%.

Management

1. Painful crises. Pain relief and adequate fluid intake. Blood, urine and a throat swab should be taken for culture and broad-spectrum antibiotics started. Non-steroidal anti-inflammatories may be adequate but opiates are frequently needed.

2. Anaemic crises. Urgent blood transfusion. In babies, splenic sequestration results in a sudden drop in Hb associated with a rapidly enlarging spleen. It can be life-threatening.

3. Prophylaxis. Penicillin and vaccination against pneumococcal infection. Increased folic acid requirements due to haemolysis.

4. Blood transfusion is not used routinely to treat anaemia as it may exacerbate hyperviscosity. Acute exchange transfusion may be of benefit in patients with acute stroke or severe sickle chest syndrome. Repeated transfusions to reduce the percentage of HbS may be useful before surgery, following a stroke (recurrence risk high within 3 years), or if painful crises are frequent.

Screening and antenatal diagnosis

Screening of cord blood is now routine in some areas of the UK and is recommended in all high-risk groups. Screening of parents will identify carriers. Sickle cell anaemia can now be detected by chorionic villus biopsy or amniocentesis.

Thalassaemia

Problems

- Anaemia.
- Growth failure.
- Bone marrow hyperplasia (skull bossing, maxillary overgrowth, brittle long bones).
- Hepatosplenomegaly (extramedullary erythropoiesis).
- Iron overload from repeated transfusions (cardiomyopathy, cirrhosis, haemosiderosis, skin pigmentation, endocrine failure).

Clinical features

Thalassaemia is most common amongst Mediterranean and Asian races. β-Thalassaemia is more common than α-thalassaemia and homozygotes (β-thalassaemia major) usually present in the first year of life with pallor and failure to thrive. Impaired β-chain synthesis leads to excessive production of α-chains, which are deposited in the red cells, causing haemolysis. Repeated blood transfusions are required to avoid the severe bone problems associated with compensatory bone marrow hyperplasia, but chronic iron overload results in significant morbidity and mortality in adolescence and early adult life. Heterozygotes (β-thalassaemia minor) have only mild anaemia which may be confused with iron deficiency anaemia.

In α-thalassaemia there are two pairs of genes involved. Lack of all four genes is incompatible with life (Bart's hydrops). Three-gene deficiency (HbH) is of varying severity and may result in transfusion dependency. Deficiency of one or two genes (α-thalassaemia trait) results in mild anaemia.

Investigation

- *FBC*. Hypochromic microcytic anaemia. Severe anaemia (Hb 4–8 g/dl) with reticulocytosis, target cells, and basophilic stippling is seen in β-thalassaemia major.
- *Haemoglobin electrophoresis.*

  β-Thalassaemia major: HbF 15–100%.

  β-Thalassaemia minor: $HbA_2 > 4\%$.
  $HbF > 2\%$.

  α-Thalassaemia: HbH detected in the three-gene defect, but normal electrophoresis in the one- or two-gene defect.

- *Radiology*. Typical 'hair on end' appearance of the skull in β-thalassaemia major.

Management

Patients with β-thalassaemia are transfused regularly to maintain their haemoglobin > 11 g/dl to try to suppress erythropoiesis. Desferrioxamine and other iron-chelating agents may help reduce iron overload and delay the onset of complications. Hypersplenism often necessitates splenectomy. Iron supplements are contraindicated in thalassaemia.

Further reading

Evans JPM. Practical management of sickle cell disease. *Archives of Disease in Childhood*, 1989; **64:** 1748–51.

Grundy R, Howard R, Evans J. Practical management of pain in sickling disorders. *Archives of Disease in Childhood*, 1993; **69:** 256–9.

Related topic of interest

Anaemia (p. 20)

SINGLE GENE DEFECTS

Single gene defects are associated with all 46 chromosomes, with the exception of the Y chromosome. Over 700 autosomal dominant, 500 autosomal recessive and 100 sex-linked disorders have been described. With modern molecular genetic techniques, the abnormal genes have been identified in many diseases, and DNA probes have enabled analysis of mutations and tracking of gene inheritance through families. The risks of having an affected child can now be determined much more accurately and antenatal diagnosis is available for a wide range of conditions.

Autosomal dominant (AD) inheritance

A mutant gene which is expressed in the heterozygote is dominant. Each child of an affected and a normal individual will have a 50:50 chance of being affected. AD traits affect both males and females, and the disorder can often be traced back through generations. The frequency with which new mutations occur varies, for example new mutations account for ~ 50% of cases of neurofibromatosis, but are virtually unheard of in Huntington's chorea. Expression may vary, producing variations in severity of the condition. Some AD conditions are sex-influenced (more commonly expressed in one sex), e.g. male pattern baldness.

Examples of AD inheritance
- Huntington's chorea.
- Tuberous sclerosis.
- Congenital spherocytosis.
- Neurofibromatosis.
- Myotonic dystrophy.

Autosomal recessive (AR) inheritance

AR disorders are expressed in individuals who are homozygous for the mutant gene, and can occur in both sexes. Both the parents of an affected child must be heterozygotes and are usually unaffected. With rare recessive traits the parents are often related (consanguineous). The risk of recurrence after one child with a recessive disorder is 1 in 4 for each successive child.

Examples of AR inheritance
- Majority of inborn errors of metabolism (heterozygotes may show intermediate levels of enzyme activity).
- Cystic fibrosis.
- Sickle cell disease.
- Ataxia telangiectasia.

In some conditions it is possible to be heterozygous for two mutant alleles, e.g. an individual who is heterozygous for both HbS and HbC has a haemoglobinopathy intermediate in severity between sickle cell anaemia and HbC disease.

Sex-linked recessive inheritance

There are no structural genes on the Y chromosome other than those determining sexual development so sex-linked conditions are all X-linked. Males are affected and females are carriers. A female can be affected in the following uncommon circumstances:

- In common X-linked conditions, an affected male and a female carrier can produce a daughter with two mutant alleles, e.g. red–green colour blindness.
- If an X-linked condition occurs in Turner's syndrome (XO), but this is very uncommon.
- Random inactivation of one X chromosome (lyonization) occurs in early embryogenesis and usually results in half the cells in the female having one X inactivated and half having the other X inactivated. Occasionally, by chance, the normal allele is inactivated in more cells than the mutant allele, resulting in partial expression of the mutant gene, e.g. ~ 5% of female carriers of Duchenne muscular dystrophy show some muscle weakness.

Examples of X-linked recessive conditions
- Chronic granulomatous disease.
- Haemophilia.
- Muscular dystrophy (Duchenne's, Becker's).
- Colour blindness.
- Glucose-6-phosphate dehydrogenase deficiency.

Sex-linked dominant inheritance

The pedigree pattern should not be confused with that of an AD condition. An affected male transmits the disorder to all his daughters and none of his sons. Offspring of an affected female will all have a 50:50 chance of being affected.

Examples of X-linked dominant conditions
- Vitamin D-resistant rickets.
- Oral–facial–digital sydrome.
- Incontinentia pigmenti – lethal in males.
- Ornithine carbamoyl transferase deficiency – lethal in males.

Screening for genetic disorders

Screening for a genetic disorder is indicated if the test is sensitive (few false negatives), specific (few false positives), safe, simple and inexpensive. It is also important that detection of the disease confers some benefit on the individual. Current programmes in the UK include the screening of all neonates for phenylketonuria and hypothyroidism using the Guthrie test at 6–7 days of life and the identification of carriers of haemoglobinopathies in high-risk subgroups. Identification of carriers allows genetic counselling and prenatal diagnosis (e.g. sickle cell disease). Early identification and treatment of infants with conditions such as hypothyroidism may help prevent mental retardation.

Further reading

Chapple JC. Genetic screening. In: *Prenatal Diagnosis and Screening*. Brock DJH, Rodeck CH, Ferguson-Smith MA, eds. Edinburgh: Churchill Livingstone, 1992; 579–93.

Related topics of interest

Antenatal diagnostics (p. 23)
Sickle cell anaemia and thalassaemia syndromes (p. 295)

SMALL FOR GESTATIONAL AGE, LARGE FOR GESTATIONAL AGE

A baby who is below the 10th centile for gestation is small for gestational age, and one who is above the 90th centile for gestation is large for gestational age. By definition, these groups will include normal small and large babies. Mortality and morbidity in these infants, particularly as a result of hypoglycaemia, has fallen dramatically as a result of better recognition both antenatally and post-natally.

Small for gestational age (SGA)

Problems

- Hypoglycaemia.
- Hypothermia.
- Polycythaemia.
- Jaundice.
- Birth asphyxia.
- Increased incidence of congenital abnormalities.

Aetiology

Babies who are SGA may be proportionately growth retarded with low weight, length and head circumference suggesting early onset of intrauterine growth retardation (IUGR) or there may be relative sparing of head circumference suggesting onset of growth retardation in the last few weeks of pregnancy.

1. *Proportionate IUGR*

(a) Fetal causes

- Congenital infection (CMV, toxoplasmosis, rubella).
- Chromosomal abnormalities.

(b) Maternal causes

- Severe placental insufficiency (e.g. pre-eclamptic toxaemia).
- Severe maternal disease (e.g. renal disease, SLE).
- Early toxins (e.g. alcohol).

2. *Disproportionate IUGR*

- Placental insufficiency (pre-eclampsia, twin-to-twin transfusion).
- Maternal smoking.
- Less severe maternal disease.

Management

1. *Hypoglycaemia* is common because of decreased glycogen stores. There may also be transient hyperinsulinism. It is often asymptomatic but clinical features include jitteriness, apnoea, irritability, floppiness

and convulsions. There is controversy over the definition of neonatal hypoglycaemia with values ranging from <1 mmol/l to <2.5 mmol/l given in various textbooks. The risk of neurological damage after repeated or prolonged episodes of hypoglycaemia is well recognized, so careful monitoring and management with early feeding, and i.v. dextrose if needed, are vital.

2. *Hypothermia* is common owing to the large surface area to volume ratio and poor fat stores. Hypothermia may exacerbate hypoglycaemia so it is important to dry the baby quickly after birth and ensure adequate wrapping. Overhead heaters and incubators are sometimes necessary.

3. *Polycythaemia* results from poor placental oxygen transfer (placental insufficiency). Blood viscosity increases exponentially when the venous haematocrit is > 65%, and this can exacerbate respiratory problems and hypoglycaemia, and may lead to long-term neurological sequelae. The precise level of haematocrit which warrants treatment is controversial, but if the baby is symptomatic (e.g. respiratory distress, irrritability, hypoglycaemia) or if the haematocrit rises to 70%, dilutional exchange should be performed. There may be an associated thrombocytopenia.

4. *Jaundice.* Increased red cell breakdown due to polycythaemia results in increased bilirubin levels. Phototherapy with or without exchange transfusion is frequently needed.

5. *Birth asphyxia.* SGA babies are at increased risk of intrapartum asphyxia because of the poor glycogen and fat stores resulting in poor ability to maintain anaerobic metabolism. There is also an increased risk of meconium aspiration.

Large for gestational age (LGA)

Problems

- Obstructed labour – shoulder dystocia.
- Birth trauma and nerve palsies.
- Birth asphyxia,
- Hypoglycaemia.

Aetiology

- Constitutionally large.
- Maternal insulin-dependent diabetes (IDDM including gestational diabetes).

- Severe rhesus disease.
- Beckwith–Wiedemann syndrome.
- Sotos' syndrome.

Infant of a diabetic mother

Problems

- LGA.
- Hypoglycaemia.
- Increased incidence of congenital abnormalities, particularly congenital heart disease, vertebral anomalies, caudal regression syndrome (e.g. sacral agenesis).
- Respiratory distress.
- Polycythaemia and jaundice.
- Hypocalcaemia.
- Cardiomyopathy.

IDDM in pregnancy increases the risk of early miscarriage, pre-eclampsia, polyhydramnios, intrauterine death and preterm labour. It is the commonest cause of a large baby (macrosomia). Congenital abnormalities are more frequent, occurring in 5–10% of infants of IDDM mothers.

Hypoglycaemia is common owing to transient hyperinsulinism (hyperglycaemia *in utero* stimulates hyperplasia of the islets of Langerhans) and reduced glucagon response to hypoglycaemia. Early feeding with regular blood sugar monitoring is essential, and an i.v. dextrose infusion may be needed.

Respiratory distress may occur even in term babies as a result of inhibition of surfactant production by hyperglycaemia *in utero*. Hypertrophic cardiomyopathy may be detected by echocardiography, with thickening of the interventricular septum, which resolves over the first year of life and only rarely causes symptoms. Other problems which are seen more frequently in these infants include polycythaemia, jaundice and hypocalcaemia.

Good diabetic control during pregnancy reduces the risk of many of these problems, but the evidence that it reduces the risk of congenital abnormalities remains to be proven. Perinatal mortality in this group is about 4%, and half of these deaths are due to major congenital abnormalities.

Beckwith–Wiedemann syndrome

This uncommon syndrome is of unknown aetiology. Features include gigantism, macroglossia, visceromegaly, hyperinsulinaemic hypoglycaemia, transverse creases on the ear lobes, and umbilical hernia or exomphalos. Symptomatic hypoglycaemia occurs in 50% of cases and may be transient, or occasionally prolonged and severe. The mental deficiency which has been a recognized part of the syndrome is probably secondary to repeated hypoglycaemia, so prompt recognition and treatment is essential. There is an increased risk of hemihypertrophy and of tumours such as Wilms' tumour and hepatoblastoma.

Further reading

Chiswick ML. Intrauterine growth retardation. *BMJ,* 1985; **291:** 845–8.

Related topics of interest

Hypoglycaemia (p. 180)
Neonatal jaundice (p. 231)
Neonatal respiratory distress (p. 235)

SOLID TUMOURS

Brain and spinal cord tumours are the most common solid tumours, accounting for 25% of all childhood malignancy (~ 300 new cases per year in the UK). The two most common extracranial solid tumours are neuroblastoma and Wilms' tumour, each with about 70 new cases per year.

Brain and spinal cord tumours

The majority of CNS tumours arise from two histological cell types: glial cells (astrocytomas and ependymomas) and neuroectodermal cells (primitive neuroectodermal tumours and medulloblastomas). They have a high potential for local recurrence and neuroaxial dissemination but distant metastases are rare.

Presentation

- Headaches.
- Early morning vomiting.
- Focal neurological signs, e.g. cerebellar signs with posterior fossa tumour.
- Increasing head circumference (infants).
- Convulsions.
- Endocrine disturbance, e.g. precocious puberty, diencephalic syndrome.

Investigation

- *Radiology.* CT scan or MRI. Fifty per cent arise in the posterior fossa.
- *Histology.* Stereotactic biopsy if unresectable. Fifty per cent are astrocytomas. Biological markers such as p53 expression may help in predicting prognosis but are still research investigations at present.

Management

Acutely raised intracranial pressure is treated with mannitol and dexamethasone. A ventriculoperitoneal shunt is sometimes needed. Surgical resection and craniospinal radiotherapy are the mainstays of treatment. Chemotherapy has been shown to shrink some tumours, particularly those of neuroectodermal origin, and may delay progression in others. Radiotherapy is particularly damaging to the infant brain so chemotherapy is currently used in this age group.

Outcome

Overall there is only a 50% chance of 5-year survival. Low-grade astrocytomas which are completely resected carry the best prognosis and high-grade gliomas (glioblastoma multiforme) are associated with a very poor prognosis, with the majority of affected infants dying within months. Late effects of the treatment include the neuroendocrine and neuropsychological effects of radiotherapy.

Neuroblastoma

This is a malignant tumour derived from the sympathetic nervous system. The peak age of incidence is 2 years; it is rare over 5 years and is slightly more common in boys (1.4M:1F). In one-third the primary tumour arises from the adrenal, and the majority have metastases at presentation (bone, bone marrow, lymph nodes, less commonly liver, skin). Prognosis is inversely related to age at presentation. Under the age of 1 year the tumour behaves very differently and may even regress spontaneously. Despite even advanced disease, the prognosis in this age group is usually good.

Presentation

- Abdominal mass (> 50%).
- Bone marrow infiltration. Anaemia, thrombocytopenia, bone pain.
- Non-specific malaise.
- Local invasion and compression, e.g. proptosis in head and neck disease, venous compression in thoracic disease.
- Spinal cord compression. A 'dumb-bell' tumour grows through the intervertebral foramina and has retroperitoneal and intraspinal components.
- Metabolic effects. High levels of catecholamines and vasoactive intestinal peptides cause sweating, pallor and diarrhoea.
- Hepatomegaly.
- Bluish skin nodules.

Investigation

- *Urinary catecholamine metabolites.* Raised vanillylmandelic acid (VMA) and homovanillic acid (HVA) on a spot urine specimen.
- *Radiology.* Abdominal US, CT scan or MRI (chest, abdomen, ± head and neck). The tumour often contains flecks of calcification.
- *Tumour histology and cytogenetics.* More differentiated tumours have a better prognosis. Structural and numerical abnormalities of chromosome 1 and N-*myc* amplification (cellular oncogene) are associated with more advanced disease.
- *Bone marrow examination.*
- *Radionucleotide bone scan.*
- *Meta-iodobenzylguanidine (MIBG) scan.* An ^{131}I-labelled guanethine analogue which is taken up by > 90% of neuroblastomas.
- *Serum ferritin.* Raised ferritin associated with more advanced disease and worse prognosis.

Management

If the tumour is localized and completely resected no further treatment may be needed. However, the majority require adjuvant chemotherapy with or without radiotherapy. Very high-dose chemotherapy (megatherapy) with autologous bone marrow rescue and targeted radiotherapy using MIBG are used in some centres.

Outcome

Localized, completely resectable disease (stage 1) has a good prognosis, with more than 90% 5-year survival, but these children account for less than 5% of the total. More than 60% of children have disseminated disease at diagnosis (stage 4). Five-year survival is ~ 60% under the age of 1 year but is only ~ 20% over 1 year, prognosis worsening with increasing age. Stage 4S is a special group under the age of 1 year, who have a localized primary tumour with dissemination to liver, skin and bone marrow. This group has more than 80% 5-year survival. Mass screening of infants for raised urinary catecholamines, to detect early neuroblastomas, has not been widely accepted since many of the tumours detected have an intrinsically good prognosis and may even regress spontaneously.

Wilms' tumour (nephroblastoma)

This tumour arises from embryonal cells within the kidney (nephrogenic rests). The majority present under the age of 5 years, boys being affected slightly more commonly than girls. The left kidney is affected more often than the right and 5–10% of patients have bilateral tumours. Wilms' tumour is associated with aniridia, hemihypertrophy, Beckwith–Wiedemann syndrome and genitourinary abnormalities.

Abnormalities of chromosome 11 have been identified in some children. Staging depends on whether there is local extension beyond the renal capsule, involvement of local lymph nodes or distant lung metastases.

Presentation

- Asymptomatic renal mass (90%).
- Haematuria – rare.

Investigation

- *Radiology.* Abdominal US and CT scan. Chest radiography to identify lung metastases.
- *Tumour histology.* Unfavourable features include areas of anaplasia.

Management

Nephrectomy is performed either as a primary procedure or following shrinkage of the tumour by chemotherapy. The duration and intensity of post-operative chemotherapy depends on the tumour stage and histological features. Radiotherapy is used if the tumour is not limited to the kidney.

Outcome

Overall there is more than 85% 5-year survival. Age over 5 years, advanced stage and unfavourable histology are worse prognostic features, but even these children have a better than 60% chance of cure.

Further reading

Murphy SB, Cohn SL, Craft AW *et al*. Do children benefit from mass screening for neuroblastoma? *Lancet*, 1991; **337:** 344–6.

Related topic of interest

Malignancy in childhood (p. 217)

STRIDOR

Stridor is a continuous harsh noise produced by obstruction to breathing in the larynx or trachea. Inspiratory stridor usually predominates, but obstruction in the subglottic area or trachea may produce a soft expiratory stridor. Stridor may arise from an acute condition, or from a chronic persistent cause with recurrent symptoms. The site of the obstruction may be supraglottic, glottic or subglottic (tracheal).

Problems

- Hypoxia – acute or chronic.
- Acute upper airways obstruction.
- Failure to thrive.
- Feeding problems.
- Infective exacerbations.

Acute stridor

Hypoxia results from the obstruction, leading to bronchospasm, pooling of secretions and patchy lung consolidation. However, intense respiratory muscle effort maintains tidal volume until the point of exhaustion. Increasing tachycardia, tachypnoea, restlessness or apathy indicates the need for urgent intervention to maintain the airway. Pulse oximetry or arterial blood gas estimation may be of little value when the clinical signs indicate impending obstruction.

Acute laryngotracheobronchitis ('croup')

This is the most common cause of acute stridor, commonly occurring during the second year of life, and is usually caused by influenza, parainfluenza and respiratory syncytial viruses.

Mild coryzal symptoms are followed by a barking cough, hoarse voice and inspiratory stridor. Symptoms are often worse at night. Steam is often helpful, although there is little supportive scientific evidence. Intubation may be needed if airway obstruction is severe. There is little indication for the routine use of steroids, but nebulized adrenaline may give temporary relief in severe cases (1 ml of 1 in 1000 diluted with saline) if full monitoring and resuscitation are available.

Acute epiglottitis

The usual pathogen is *Haemophilus influenzae* type B, occasionally the cause is streptococcal or viral. There is gross swelling of the epiglottis and aryepiglottic folds, causing a rapid onset of drooling, fever, muffled voice and quiet inspiratory stridor. Cough is not prominent. Typically the child sits forward to relieve the obstruction. Throat examination and unnecessary upset to the child should be avoided – complete respiratory obstruction may result.

Expert anaesthetic and ENT assistance should be sought immediately, and laryngoscopy and intubation performed. Intravenous antibiotics may be given once the airway is secure – cefotaxime is appropriate. If *Haemophilus influenzae* is the cause, rifampicin prophylaxis should be given to the index case and to household contacts.

Foreign body

There may be a history of choking or aspiration. Stridor has a variable quality, depending on site and mobility of object. Urgent rigid bronchoscopy under general anaesthetic is required.

Retropharyngeal abscess (less common)

This is often caused by anaerobic or Gram-negative organisms and may follow bacterial pharyngitis or trauma in the young child. Soft stridor, neck extension, cervical lymphadenopathy and dysphagia are features. Treatment should include expert airway management, surgical drainage and broad-spectrum intravenous antibiotics.

Other causes of acute stridor are less common and include bacterial tracheitis, tonsillar obstruction (quinsy, severe infectious mononucleosis), angioneurotic oedema, diphtheria and thermal, chemical or mechanical trauma to the upper airway.

Chronic stridor

The history of a chronic stridor may indicate that it dated from birth or the early neonatal period. Its quality and phase (inspiratory or expiratory), the nature of the cry and the presence of associated symptoms such as feeding difficulties or dyspnoea may give diagnostic clues. There may be persistent sternal retraction, or a Harrison's sulcus if obstruction is chronic, and failure to thrive may reflect chronically increased respiratory effort.

Congenital laryngeal stridor (laryngomalacia, infantile larynx)

There is an elongated epiglottis, a relatively small larynx and redundant epiglottic folds. There may be associated micrognathia. Inspiratory stridor presents within 4 weeks of birth, worsening by 3–6 months. Twenty per cent also have an expiratory stridor. Stridor is worse in the supine position, the cry and cough are normal and the child is usually thriving. A presumptive diagnosis is often made, but floppy aryepiglottic folds are seen on flexible bronchoscopy. Spontaneous improvement by 2–4 years is the rule.

Subglottic stenosis

There may be a congenital thickening of subglottic tissues or true cords, congenital malformation of the cricoid cartilage or acquired stenosis from prolonged endotracheal intubation.

There is both inspiratory and expiratory stridor, worsened by intercurrent infection. Stridor usually improves with growth.

Subglottic haemangioma

From 1 to 3 months of age, there is variable inspiratory and expiratory stridor which worsens with crying. Fifty per cent have cutaneous haemangiomas. Tracheostomy is often required; cautery, sclerosants, steroids and laser therapy may all have a role in treatment.

Vocal cord palsy

This is often associated with severe CNS deformity or major cardiac lesions. Inspiratory stridor, respiratory distress and feeding difficulties are common features.

Other causes

Laryngeal webs, cysts, clefts and papillomas are infrequent causes of stridor.

Tracheal compression, e.g. a vascular ring, is also uncommon. A double aortic arch or abnormally placed major artery may cause soft inspiratory stridor with expiratory wheeze. There may be occasional dysphagia and regurgitation.

Investigation

- Chest radiography and lateral neck radiography of soft tissue.
- Barium swallow.
- Angiography.
- CT scan and MRI scan.
- Direct laryngoscopy and bronchoscopy.

Management

The approach to management will depend on the causative lesion, e.g surgical relief of extrinsic compression from a vascular ring, awaiting spontaneous resolution in laryngomalacia. Tracheostomy itself carries risks of scarring, stenosis and obstruction, but may be necessary for temporary or permanent relief of airway obstruction.

Further reading

Couriel JM. Management of croup. *Archives of Disease in Childhood,* 1988; **63:** 1305–8.

Kilham HA, McEniery JA. Acute upper airway obstruction. *Current Paediatrics,* 1991; **1(1):** 17–25.

MacFadyen U. Chronic upper airways obstruction. *Current Paediatrics,* 1991; **1(1):** 26–9.

Related topics of interest

Lower respiratory tract infection (p. 211)
Upper respiratory tract infection (p. 319)

SUDDEN INFANT DEATH SYNDROME

Sudden infant death syndrome (SIDS) is defined as the sudden death of an infant or young child which is unexpected by history, and in which a thorough post-mortem examination fails to demonstrate an adequate cause of death. It is clear that there is no single underlying cause, but that numerous factors may coincide to cause death in susceptible babies, and that the final common event is the cessation of respiration prior to circulatory arrest. Until 1988, the UK prevalence had been a constant 2 per thousand, but this figure has been declining. This appears to coincide with the national campaign to avoid known risk factors in infant care practices (see below), particularly prone sleeping.

Epidemiological risk factors

SIDS is rare before 1 month and after 6 months of age, with a peak at 2–3 months. SIDS occurs with increased frequency in preterm and low birth weight babies, and in those from multiple births. SIDS deaths are more common during winter months. There is a three- to sevenfold increased risk for subsequent siblings. Babies who have bronchopulmonary dysplasia or who have suffered an apparent life-threatening event are also at increased risk. SIDS is more common in babies born to young mothers, those of high parity, those who take opiates and in those for whom housing, employment and financial status is poor. Maternal smoking has been established as an independent and potentially avoidable risk factor.

Aetiological factors

1. Sleeping position. Evidence from many studies has shown that the prone sleeping position is an independent risk factor for SIDS. The results of population-based intervention studies have shown that a fall in the prevalence of prone sleeping has been accompanied by a fall in the number of sudden infant deaths. A national campaign was launched in the UK in 1991, and was supported by Department of Health 'Back to Sleep' advice.

2. Infection. The role of infection in SIDS is unclear and, although many parents recall the presence of minor symptoms of illness before death, there seems to be no significant difference between victims and controls. In a small number of cases, fulminant bacterial infection will be a factor, although interpretation of both bacterial and viral isolates from dead infants is difficult. However, overwrapping the baby during a viral infection increases the risk of SIDS.

3. Overwrapping. Warm sleeping environments may be harmful to certain vulnerable babies. There may be combined factors such as excessive wrapping with heavy duvets or quilts, high ambient temperature and possibly the direct heat of an adult body or electric blanket.

4. *Apnoea.* There is an increased frequency of apnoeic pauses in SIDS victims. This may represent a failure of arousal or an abnormal ventilatory response to a period of relative hypoxia or hypercarbia.

5. *Inborn errors of metabolism.* Very few of these disorders are likely to cause SIDS, although the true incidence is unknown. There is evidence that medium-chain acyl CoA dehydrogenase (MCAD) deficiency may be responsible in some cases, and specific investigation for metabolic disease is recommended for those in whom there is parental consanguinity, previous SID of sibling, previous unexplained illness with drowsiness, hypoglycaemia, etc., or if post-mortem changes of fat or glycogen deposition are present.

6. *Feeding and child-care practices.* There is a lack of evidence for the protective effect of breast-feeding, although this should be promoted for numerous other reasons, and should be studied further in future research. National variations in incidence of SIDS are wide, and in communities in which babies are rarely left alone to sleep SIDS is rare.

Management

The most experienced available paediatrician should see the parents as soon as possible, explaining the necessary early involvement of the coroner and police and the need for a post-mortem. Ample time should be allowed for parents to hold and cuddle their baby. Continued follow-up appointments should be offered, including a discussion of the post-mortem findings. Counselling is particularly important if there is a subsequent sibling, and parents may seek the reassurance of an apnoea monitor. There is no evidence that they prevent SIDS, however, and they should always be accompanied by practical and written training of cardiopulmonary resuscitation, the care of the next infant scheme (CONI) provides important support.

Primary prevention

Early child health surveillance programmes should include advice to place the baby supine (or on the side with the lower arm well in front of the baby to avoid rolling prone), to avoid smoking, and to avoid overwrapping the baby.

Further reading

Green A. Biochemical screening in newborn siblings of cases of SIDS. *Archives of Disease in Childhood*, 1993; **68:** 793–6.

Wigfield RE, Fleming PJ, Berry PJ, *et al.* Can the fall in Avon's sudden infant death rate be explained by the observed sleeping position changes? *BMJ,* 1992; **304:** 282–3.

Related topics of interest

Cardiopulmonary resuscitation (p. 52)
Inborn errors of metabolism (p. 191)

THYROID DISORDERS

Thyroid-stimulating hormone (TSH) controls the synthesis and release of thyroxine (T_4) by the thyroid gland. TSH secretion by the pituitary is under the control of hypothalamic thyrotrophin-releasing hormone (TRH), and is inhibited by somatostatin and dopamine. Triiodothyronine (T_3) is more metabolically active than T_4 and some is synthesized by the thyroid gland, but most is produced by conversion of T_4 to T_3 in peripheral tissues. T_4 is bound in plasma to thyroid-binding globulin (TBG), and T_3 is bound to TBG and albumin. The metabolically active free T_4 and T_3 concentrations can now be measured directly by radioimmunoassay and account for only a small percentage of the total plasma concentrations. T_3 and T_4 modulate many physiological processes including thermogenesis and growth factor production. They also potentiate the actions of catecholamines and have widespread functions in the development of the CNS.

Hypothyroidism

Aetiology

1. *Primary hypothyroidism (T_4 low, TSH high).*
- Congenital hypothyroidism (thyroid dysgenesis or, less commonly, disorders of hormone synthesis).
- Iodine deficiency.
- Goitrogens.
- Autoimmune thyroiditis.
- Post irradiation or chemotherapy.

2. *Secondary hypothyroidism (T_4 low, TSH normal or low).*
- TSH deficiency (congenital or acquired).

3. *Tertiary hypothyroidism (T_4 low, TSH low).*
- TRH deficiency.

4. *TBG deficiency.*
- Total T_4 low but free T_4 normal – clinically euthyroid.

Clinical features

1. *Congenital hypothyroidism* occurs in 1 in 4000 live births, and screening for a raised TSH is now included in the Guthrie test. A raised TSH indicates primary hypothyroidism, but secondary and tertiary hypothyroidism (accounting for less than 10% of cases) will not be detected by this screening method as the TSH is normal or low. Infants are usually asymptomatic at birth, with features such as umbilical hernia, prolonged jaundice, large tongue, feeding difficulties, constipation and lethargy developing over 6–12 weeks if the baby is untreated. Growth failure will become increasingly obvious over the first year. Congenital TRH or TSH deficiency may occur as a component of

panhypopituitarism (hypoglycaemia, micropenis, midline cranial defects).

2. *Acquired hypothyroidism* can occur at any age, and the onset of clinical features, such as weight gain, short stature, constipation and mental slowness, is usually insidious. A goitre may develop, tendon jerks are slow-relaxing and the bone age is delayed. Other presenting features include isolated breast development or large testes, without other signs of puberty, since TRH also stimulates FSH and prolactin secretion. Autoimmune thyroiditis may present with hypothyroidism, thyrotoxicosis or compensated hypothyroidism (raised TSH, normal T_4 – clinically euthyroid). It is more common in girls, may be familial and is associated with other autoimmune conditions. Thyroid disease is more common in Down's, Turner's and Noonan's syndromes.

Investigation

- *Free T_4 and T_3.*
- *TSH.*
- *TRH stimulation test.* Useful if the TSH level is borderline. There is an exaggerated TSH response in primary hypothyroidism, a negligible TSH response in secondary hypothyroidism and a TSH rise in tertiary hypothyroidism.
- *Thyroid antibodies*, e.g. antithyroglobulin, antimicrosomal antibodies.
- *Radiology.* Delayed bone maturation, epiphyseal dysgenesis (e.g. femoral heads)
- *Radioisotope thyroid scan.* Not essential but may identify ectopic thyroid tissue. The perchlorate discharge test is used to identify enzyme defects.

Management

Hypothyroidism should be treated with oral thyroxine and the dose adjusted to maintain a near normal TSH and a satisfactory growth velocity. Primary hypothyroidism is sometimes transient, but all babies with a raised TSH need treatment for at least the first year of life. For the majority of patients (congenital and acquired) treatment is life-long. Early treatment of congenital hypothyroidism has led to a marked reduction in mental handicap, but mild learning difficulties usually persist. Parents should be warned that the behaviour of older children often deteriorates when they are started on thyroxine, as they emerge from hypothyroid docility.

A persistently raised TSH is a risk factor for thyroid carcinoma so should be treated with thyroxine even if the child is euthyroid. Hypothyroidism is a common late effect of cancer treatment, and these children are particularly at risk of second tumours.

Hyperthyroidism

Hyperthyroidism in childhood is usually autoimmune. It is more common in girls and may be familial. Presenting features include behavioural problems, poor school performance, accelerated growth and weight loss in spite of an increased appetite. Clinical signs include mild exophthalmos, tachycardia, tremor and lid-lag. There may be a thyroid bruit or a goitre. Investigations show a high T_4 and T_3, a low or occasionally normal TSH and no response to TRH stimulation. Thyroid autoantibodies are usually positive. Treatment is with carbimazole or propylthiouracil, and propanolol is used acutely to control the β-adrenergic symptoms. Approximately half of childhood cases will remit spontaneously within 2–4 years, and many will become hypothyroid. Partial thyroidectomy is an option if hyperthyroidism persists. Neonatal thyrotoxicosis is a transient condition resulting from placental transfer of thyroid-stimulating antibodies from a thyrotoxic mother. Symptoms are often severe but can usually be controlled with propanolol. Potassium iodide or carbimazole is used to suppress thyroid activity in severely affected babies. Treatment can be withdrawn slowly after 2–3 months.

Further reading

Brook CGD. The thyroid gland. In: *A Guide to the Practice of Paediatric Endocrinology*. Cambridge University Press, 1993: 83–96.

Grant DB. Monitoring TSH concentrations during treatment for congenital hypothyroidism. *Archives of Disease in Childhood,* 1991; **66:** 669–71.

Related topic of interest

Growth – short and tall stature (p. 149)

UPPER RESPIRATORY TRACT INFECTION

Acute infection of the upper respiratory tract affects the average child approximately four times per year. The ears, nose, tonsils, pharynx or sinuses may be affected, or any combination of these. Viruses are most commonly responsible, although serious bacterial infection may also occur. Certain epidemiological factors, such as parental smoking and overcrowding, may increase the rate or severity of infection.

Problems

- Recurrent infection and upset.
- Acute feeding disturbance.
- Secondary bacterial infection.
- Acute and chronic upper airway obstruction.
- Febrile convulsions.

Common cold (coryza, nasopharyngitis)

Rhinovirus, adenovirus, respiratory syncytial virus, influenza and parainfluenza are commonly responsible. Sneezing, sore throat, nasal obstruction and discharge occurs after a 2- to 3-day incubation period and may persist for up to 2 weeks. Nasal obstruction may cause feeding difficulties in the younger infant; saline nose drops may help clear mucus. Complications include direct viral invasion of the middle ear, sinuses or lower respiratory tract and secondary bacterial infection. Symptomatic treatment should be given with attention to fluid balance and paracetamol if necessary. There is little evidence to suggest that decongestants or antihistamines are of benefit.

Pharyngitis and tonsillitis

Adenovirus, parainfluenza and influenza virus, Epstein–Barr virus, coxsackie- and echoviruses are responsible in the majority of cases. Group A β-haemolytic streptococcus and other bacteria are causative in a significant number of cases. Sore throat, fever and malaise is accompanied by a red throat or tonsils, with exudate and cervical lymphadenopathy. Clinically it is difficult to distinguish between viral and streptococcal aetiology, and since the streptococcus may lead to sequelae of rheumatic fever or acute glomerulonephritis the decision to withold treatment may be difficult. Treatment with oral penicillin for 10 days should eradicate streptococcus. A long-acting intramuscular penicillin can be used if compliance or absorption is in doubt. Confirmation of a bacterial cause may be obtained by a throat swab if practical circumstances permit. Patients with a past history of rheumatic fever or nephritis should always be treated.

Acute otitis media

The aetiology is viral in many cases, e.g. respiratory syncytial virus, adenovirus, influenza, but a significant number of bacterial pathogens may also be responsible, including *Streptococcus pneumoniae, Haemophilus influenzae*, group A β-haemolytic streptococcus. Severe otalgia is accompanied by fever, irritability and a bulging tympanic membrane with loss of its light reflex. Purulent discharge may appear if the drum ruptures. Complications include chronic suppurative infection and secretory otitis media with effusion. Meningitis, mastoiditis and cerebral abcess may rarely complicate an acute bacterial episode. Treatment with antibiotics (penicillin, amoxycillin, erythromycin or trimethoprim) is appropriate, since clinical distinction between viral and bacterial otitis media is difficult. Paracetamol may be given for fever and pain.

Sinusitis

Infection of paranasal sinuses may arise from spread of a viral infection during an acute coryzal infection, or by secondary bacterial infection. Facial pain, fever, headache and a purulent nasal discharge may be seen, with localized tenderness over the affected sinus. Complications include periorbital cellulitis, cavernous sinus thrombosis and cerebral abscess. Antibiotics should be given if bacterial infection is suspected, with symptomatic treatment of pain and fever. Surgical drainage is rarely required. If sinusitis is chronic, conditions such as allergy, cystic fibrosis, ciliary dyskinesia and immunodeficiency should be considered.

Acute laryngotracheo-bronchitis, acute epiglottitis

See related topic, Stridor, p. 310.

Adenotonsillectomy

This surgery is widely performed, although there has been little objective evidence in the past to show a definite beneficial effect for many children, and controversy continues. For children with recurrent tonsillitis in particular, it is important to apply careful selection criteria, which include an assessment of the frequency of documented tonsillitis episodes and the amount of education missed. However, indications for adenotonsillectomy may be considered as:

- Obstructive sleep apnoea. Tonsillar and adenoidal hypertrophy may contribute to chronic upper airway obstruction, with the development of right ventricular hypertrophy and, eventually, cor pulmonale.

- Investigation of malignancy. Excision biopsy of the tonsil may rarely be indicated.
- Peritonsillar abscess. Tonsillectomy is usually advised because of the risks of rupture with aspiration, upper airway obstruction, or involvement of the retropharyngeal tissues.
- Recurrent tonsillitis. If significant numbers of school days are missed (approximately 3 weeks) during two successive years because of well-documented episodes of tonsillitis each year, tonsillectomy may be of benefit. Adeno-tonsillectomy may also be recommended for recurrent otitis media which is associated with tonsillitis.

Further reading

Maw AR. Developments in ear, nose and throat surgery. In: David TJ, ed. *Recent Advances in Paediatrics,* Churchill Livingstone, Vol. 9. Edinburgh: 1991: 93–108.

Related topics of interest

Lower respiratory tract infection (p. 211)
Stridor (p. 310)

URINARY TRACT INFECTION

Urinary tract infection (UTI) is common in infants and children. Infection in the presence of vesicoureteric reflux of urine is associated with renal cortical scarring (reflux nephropathy), particularly in the first year of life. This can have significant long-term implications, underlying up to 20% of cases of chronic renal failure in children and being a significant cause of end-stage renal failure in young adults. It is important that a UTI is proven on urine culture and not just treated on the basis of symptoms, as the diagnosis should initiate thorough radiological investigation of the renal tract.

Problems

- Enuresis.
- Failure to thrive.
- Septicaemia.
- Reflux nephropathy (hypertension, chronic and end-stage renal failure, increased incidence of pre-eclamptic toxaemia in pregnancy).

Aetiology

Most cases are caused by normal bowel commensal organisms which ascend the urethra (> 90% of first infections are due to *E. coli*). In the neonatal period, infection may be haematogenously acquired and bacteraemia is present in > 60%.

Predisposing factors include incomplete emptying of the bladder (neuropathic bladder, mechanical outflow obstruction, e.g. ureterocele), constipation and poor fluid intake.

Clinical features

In the neonatal period UTI is more common in males (2M:1F). Presenting features include irritability, refusal to feed, vomiting, failure to thrive and exacerbation of jaundice. If bacteraemic, the baby will be toxic and extremely unwell. In older children UTIs are more common in girls (prevalence 1–2% of schoolgirls, 0.2% of boys). The pre-school child may present with non-specific diarrhoea and vomiting, poor weight gain, fever or malaise. More specific symptoms include frequency, dysuria, enuresis, fever with or without haematuria, loin pain.

Frequency and dysuria may also be caused by irritation from nylon underwear, bubblebaths or perineal *Candida* in the absence of a UTI.

Examination should include blood pressure, palpation of the kidneys, lower limb reflexes and examination of the lumbosacral spine.

Investigation

- Urine microscopy, culture and sensitivity. Reliability depends on the care of collection. The collection method in an older child is a mid-stream urine (MSU) sample and in a young child/infant bag urine, clean catch or suprapubic aspirate (SPA). Pyuria suggests infection but is not diagnostic. Organisms may be seen on microscopy. Significant bacteriuria is defined as a pure growth of more than 105 organisms per ml of clean, freshly passed urine, or any growth from an SPA. Two urine specimens should be collected unless an SPA is performed.

- Blood cultures.
- FBC.
- U&E, creatinine.

Management

If a UTI is suspected appropriate antibacterial therapy should be commenced once the urine specimens have been collected. Antibiotics should be given intravenously if the child is vomiting or is toxic and unwell. High fluid intake should be encouraged. After a short course of antibiotics the urine should be re-examined to ensure that the infection has cleared. Low-dose prophylactic antibiotics should be continued at least until further investigation of the urinary tract has been completed

Further investigation

It is essential that all young children with a proven UTI are investigated radiologically to detect scars and vesicoureteric reflux (VUR) so that prophylaxis against further infections can be given in an attempt to prevent reflux nephropathy. VUR is found in 30–40% of children presenting with a UTI, and the prevalence of VUR is as high as 1 in 250 children with a high familial incidence. More than 80% will resolve spontaneously with age, but there is a high incidence of cortical scarring associated with VUR, particularly in children < 1 year, suggesting that susceptibility to scarring is greatest in infants. There is no consensus as to what investigations should be performed, but one suggested approach following a first proven UTI is:

1. *Infants under 1 year.* US and dimercaptosuccinic acid (DMSA) isotope scans to detect renal scars and a micturating cystourethrogram (MCUG) to detect VUR. DMSA scans should not be performed within 4 weeks of a UTI as they may detect transient areas of abnormality which do not develop into scars on long-term follow-up.

2. *Age 1–5 years.* US and DMSA scans. If scars are detected, or if there are recurrent UTIs proceed to an MCUG (this investigation involves catheterization and is unpleasant for an older child).

3. Age more than 5 years. US only. Other investigations are indicated only if US is abnormal.

Prevention of reflux nephropathy

Simple measures which may help prevent UTIs include a good fluid intake, regular voiding of the bladder and correction of constipation.

If VUR is detected, prophylactic low-dose antibiotics should be given to prevent further UTIs, as evidence suggests that reflux of sterile urine does not cause scarring. New scars probably do not develop after the age of 5 years, but some centres continue prophylactic antibiotics to the age of 10 years. Compliance becomes a problem the longer antibiotics are continued. Prophylactic antibiotics may have a place in the treatment of older children with recurrent symptomatic UTIs. A small number of children with severe reflux, recurrent UTIs despite prophylactic antibiotics and evidence of new scars developing, proceed to surgery such as reimplantation of the ureters or silicone injection of the VU junctions.

Further reading

Guidelines for the management of acute urinary tract infection in childhood. Report of a working party. *Journal of the Royal College of Physicians of London,* 1991; **25(1):** 36–42.

Smellie JM, Normand ICS, Prescod N, Edwards D. Development of new renal scars: a collaborative study. *BMJ,* **290(6486):** 1985; 1957–60.

White RHR. Vesico-ureteric reflux and renal scarring. *Archives of Disease in Childhood,* 1989; **64:** 407–12.

Related topics of interest

Chronic renal failure (p. 75)
Haematuria (p. 153)

VISION

Normal vision depends on precise motor and sensory optic functions in addition to numerous central interactions which interpret the information received. Abnormalities of vision in childhood may be minor, e.g. mild refractive errors, or major, resulting in severe visual impairment and disability.

Refractive errors

These are the most common visual problem in childhood. Images are not formed precisely on the retina. If the eye is otherwise healthy, errors may be corrected by wearing lenses. Hypermetropia is uncommon. Myopia frequently presents during school years and there is a familial tendency. Astigmatism produces a distorted image because the corneal curvature is non-uniform, but may be corrected by appropriate lenses.

Squints

Squints (strabismus) affect 3.5% of children. A squint is an abnormal alignment of the eyes which prevents the occurrence of single binocular vision. Pseudosquints may occur if the child has wide epicanthic folds. Squints are usually concomitant in children, i.e. the angle of squint is the same in all directions of gaze. If the squint is incomitant (paralytic), the angle of squint alters with different positions of gaze, or with changes of fixation. Manifest squints are readily apparent at examination and are demonstrated by the cover test. Latent squints are detected at times of tiredness or stress, or by specific clinical examination (alternate cover test). Early detection of squints is important for the diagnosis of treatable causes, e.g. cataracts, and to prevent the development of amblyopia.

Amblyopia

Amblyopia results when the brain has suppressed or failed to develop the ability to perceive a detailed image from one eye and is usually defined as a reduction of at least two lines on the Snellen chart. Loss of acuity may be severe, and prevention by successful detection and treatment of squints is of great importance.

Defects of colour vision

These occur in 8% of boys and 0.5% of girls, and chiefly affect red–green perception. Many types are X-linked recessive. Detection is important for careers guidance (certain occupations barred) but may be relevant if classroom methods are colour coded.

Serious impairment of vision

One in 2500 children is registered blind, usually requiring educational methods which do not involve sight with a visual acuity < 6/12. Others may be partially sighted, requiring educational methods which rely on some residual vision whose acuity is < 6/60. The aetiology of such serious defects includes cataracts, optic atrophy, retinal disease (retinoblastoma, retinopathy of prematurity, chorioretinitis), congenital eye defects and severe refractive errors. Corneal ulceration is the commonest cause of blindness world-wide. Neoplasia of the optic nerve or tract may cause blindness.

History

A full history should include family history of visual problems, pregnancy and birth history. Predisposing conditions may be readily apparent, e.g. Down's syndrome, hydrocephalus. Significant post-natal events such as head injury or meningitis must also be noted. Parental concerns about the child's visual performance may be expressed, and the symptoms of nystagmus with roving eye movements, light gazing, photophobia or eye pressing may be reported. Language and motor development may be delayed.

Testing of vision

Routine vision screening in the pre-school child has been a controversial subject. Parental concerns are important at any age. Visual acuity should be tested at all assessments, and squints should be excluded by shining a light source (looking for asymmetrical reflection if there is a squint) and use of the cover and uncover test.

Screening at 6 weeks of age should include testing of pupil reactions, fixing and following, examination of the red reflex with an ophthalmoscope and exclusion of squint. Vision should be tested at school entry and at 3-yearly intervals thereafter. Concerns about vision should lead to a fuller assessment.

The Catford drum, tests of preferential looking, and the Sheridan graded balls (mounted or rolling at a distance of 3 m) may be of value in the young infant. Above 1 year of age, tests of letter matching or naming (Stycar or Sheridan–Gardiner tests) may be used, always testing each eye separately. Other tests, recently designed, e.g. the Sonksen–Silver acuity system, may prove useful. From 4 to 5 years, the Snellen chart can be used.

Tests of colour vision are the Ishihara and Sheridan–Gardiner screening tests, and also the City University test.

Videorefraction is a newer technique. Electroretinography and visually evoked potentials may be useful in the neonate or severely handicapped child.

Management

Refractive errors may be corrected by appropriate lenses. Squints may be treated by exercises, occlusion ('patching'), spectacles or surgery, aiming to restore binocular vision before amblyopia occurs. Other remediable conditions must be treated early, e.g. retinopathy of prematurity by cryotherapy, surgical removal of cataracts. For the blind child, management depends on early intervention programmes to maximize the use of the child's other special senses. Special educational provision must be made for their learning needs, and skills in learning Braille may be taught. Genetic counselling is important in many cases of blindness.

Further reading

Fielder AR. The management of squint. *Archives of Disease in Childhood,* 1989; **64:** 413–8.

Hall DMB, ed. *Health for all Children. A Programme for Child Health Surveillance.* Oxford: Oxford University Press, 1991.

Sonksen PM. The assessment of vision in the preschool child. *Archives of Disease in Childhood,* 1993; **68:** 513–6.

Related topics of interest

Antenatal diagnostics (p. 23)
Developmental assessment (p. 111)
Developmental delay and regression (p. 115)
Hearing and speech (p. 165)

VOMITING

Vomiting is a very common symptom in all paediatric age groups, ranging from the benign posseting of infancy to the sinister, effortless vomiting associated with raised intracranial pressure. Posseting is regarded as the regurgitation of small amounts of milk after feeds and does not cause failure to thrive. Regurgitation of larger amounts of liquid feed occurs in gastro-oesophageal reflux, and significant loss of calories may result. Vomiting of a true nature occurs with force, expelling most of the stomach contents with each episode.

Aetiology

The causes of vomiting are numerous, but the following categories of disease should be considered:

1. Infective. Gastroenteritis, urinary tract infection, meningitis, tonsillitis, otitis media or lower respiratory tract infection.

2. Intestinal obstruction. Intussusception, pyloric stenosis, intestinal atresias, acute appendicitis, volvulus, strangulated hernia.

3. Gastro-oesophageal reflux.

4. Intracranial pathology. Minor head injuries. Raised intracranial pressure from subdural haematoma, hydrocephalus, intracranial tumour, encephalitis, meningitis.

5. Food allergy or intolerance. Cows' milk, egg, soy, rice intolerance, coeliac disease.

6. Metabolic. Diabetic ketoacidosis, inborn errors of metabolism, uraemia, Reye's syndrome, hypercalcaemia

7. Psychological. As a feature of cyclical vomiting, or infrequently with excitement or anxiety

8. Drugs and toxins. Cytotoxic agents, antibiotics, theophylline, opiates, iron, digoxin, lead poisoning.

Gastro-oesophageal reflux (GOR)

Regurgitation and vomiting are common symptoms of gastro-oesophageal reflux in infants, often with failure to thrive. There may be associated oesophagitis, causing dysphagia, irritability, haematemesis and melaena or stricture formation. However 'silent' reflux may cause respiratory symptoms, apparent life-threatening events, or neurobehavioural symptoms, e.g. Sandifer–Sutcliffe syndrome. Twenty-four hour oesophageal pH monitoring is currently the method of choice for detecting and quantifying GOR – barium studies may miss the dynamic process of reflux, but are of value in detecting anatomical lesions, e.g. malrotation, hiatus hernia. Treatment includes positioning in a prone position, milk thickeners, and antacids initially, with the addition of the prokinetic agents (e.g. cisapride or domperidone) if there is no improvement. Dopamine antagonists such as domperidone or metoclopramide may also be used. H_2-receptor antagonists, sucralfate or omeprazole, are indicated for those with complications such as oesophagitis, and if medical therapy is unsuccessful surgical intervention (e.g Nissen fundoplication) may be required.

Hypertrophic pyloric stenosis

Presentation with projectile vomiting between the third and eighth week of life is typical. Vomiting is often infrequent initially, gradually increasing as pyloric obstruction progresses. Males are more commonly affected than females, and there is often a positive family history. Prolonged jaundice, constipation and failure to thrive may be accompanying features. On examination, there may be signs of dehydration, and a test feed shows visible waves of peristalsis moving from left to right across the abdomen, with a palpable pyloric 'tumour' in the right hypochondrium. Hypochloraemic alkalosis with hyponatraemia and hypokalaemia is typical. Diagnosis may be confirmed by US scan or barium contrast studies. Correction of electrolyte disturbances and dehydration should always precede surgery (Ramstedt's pyloromyotomy).

History

Diagnostic possibilities are often suggested by the child's age, e.g. pyloric stenosis. The onset, nature and frequency of vomiting should be established, and the presence of blood or bile staining. Enquiry should be made about associated symptoms such as fever, diarrhoea, abdominal pain, headache or urinary symptoms. If the nature or frequency of vomiting is difficult to establish from the history alone, a period of in-patient observation may be of value.

Examination

Height and weight should be plotted on the centile charts. General examination should include assessment of the state of hydration, nutrition, consciousness level and respiratory rate. There may be signs of local or systemic infection. Careful examination of the abdomen is essential, and may reveal masses, organomegaly, tenderness or distension.

Investigation

Extensive investigation is unnecessary in every case. The following tests may be indicated:

- FBC.
- U&E, calcium, creatinine, acid–base balance, glucose. Ammonia and organic acids may be indicated.
- Urine culture. Urine amino acids.
- Blood culture if there are signs of systemic infection.
- Abdominal radiography – erect and supine.
- Ultrasound scan of abdomen – specifically renal tract and pyloric area of stomach.
- Barium studies. Examine for hiatus hernia, peptic ulceration, pyloric stenosis.
- Oesophageal pH monitoring.
- Upper gastrointestinal endoscopy.
- CT, MRI imaging of brain if intracranial pathology is suspected.

Further reading

Booth IW. Silent reflux: how much do we miss? *Archives of Disease in Childhood,* 1992; **67:** 1325–7.

Milla PJ. Reflux vomiting. *Archives of Disease in Childhood,* 1990; **65:** 996–9.

Sullivan PB, Brueton MJ. Vomiting in infants and children. *Current Paediatrics,* 1991; **1(1):** 13–16.

Related topics of interest

Abdominal pain – acute (p. 1)

Gastroenteritis (p. 138)

Sudden infant death syndrome (p. 313)

Urinary tract infection (p. 322)

INDEX

Key Topics in Accident & Emergency Medicine

P. Howarth & R. Evans
respectively Royal Cornwall Hospital, Truro, UK; and Cardiff Royal Infirmary, UK

Contains essential information on the major topics associated with acute injury and sudden illness relevant to all medical and nursing staff working in Accident and Emergency and Acute Medical and Surgical Departments. The book is an ideal revision aid for the Part II Fellowship examination in Surgery or Accident and Emergency Medicine, the new Faculty examination in Emergency Medicine, or other postgraduate qualification. It is also an invaluable source of reference for anyone who deals with acute problems, including paramedics, general practitioners and nurse practitioners.

Contents

Topics cover a wide variety of conditions which commonly present to Accident and Emergency Departments. Individual topics are structured in a uniform, systematic style for ease of reference. Typically, each topic includes diagnosis and immediate management but also includes short reviews of topical subjects and current practice.

Of interest to:

Medical and nursing staff working in Emergency and Surgical Departments, particularly candidates for Fellowship or other postgraduate qualifications; paramedics; nurses and practitioners.

Paperback; 352 pages; 1-872748-67-8; 1994

Resuscitation: Key Data

M.J.A. Parr & T.M. Craft
respectively Bristol Royal Infirmary; and Royal United Hospital, Bath, UK

This pocket reference guide is an invaluable collection of essential data and treatment guidelines for the resuscitation of neonatal, paediatric and adult patients. Information is presented in a clear format which allows instant access to key data in an emergency situation.

Based on the latest recommendations from the European Resuscitation Council, the American College of Surgeons and other bodies, together with information derived from BCLS, ACLS, ATLS and PALS training courses. Treatment protocols for a broad range of conditions including trauma, burns, cardiac arrhythmias and drug overdose are clearly presented as flowcharts and decision trees. Tabulated information includes Apgar and Glasgow coma scoring, the constituents of intravenous fluids, antiarrhthmic drug doses and normal values for some of the more common investigations.

"This book is definitely not one for gathering dust on the bookshelf, but would be more at home in the white coat pocket where I have no doubt it will become well thumbed."
J. Br. Assoc. Immediate Care

"Don't just keep it in your pocket. read it, use it and keep using it."
Colin Robertson, Chairman of the UK Resuscitation Council

Contents

Part 1 Adult resuscitation: Basic life support; Advanced life support; Cardiac; Trauma; Burns; Anaphylaxis; Acute severe asthma; Hypothermia; Drug overdose. Part 2 Paediatric resuscitation: Basic life support; Advanced life support; Newborn; Infant and child; Trauma; Burns. Part 3 Normal physiological values: Biochemistry; Haematology; Coagulation; Blood gases; Conversion factors.

Of interest to:

All doctors, nurses, paramedics and practitioners trained in resuscitation.

Paperback; 96 pages; 1-872748-53-8; 1994

ORDERING DETAILS

Main address for orders

BIOS Scientific Publishers Ltd
St Thomas House, Becket Street,
Oxford OX1 1SJ, UK
Tel: +44 865 726826
Fax: +44 865 246823

Australia and New Zealand

DA Information Services
648 Whitehorse Road, Mitcham, Victoria 3132, Australia
Tel: (03) 873 4411
Fax: (03) 873 5679

India

Viva Books Private Ltd
4346/4C Ansari Road, New Delhi 110 002, India
Tel: 11 3283121
Fax: 11 3267224

Singapore and South East Asia

(Brunei, Hong Kong, Indonesia, Korea, Malaysia, the Philippines, Singapore, Taiwan, and Thailand)

Toppan Company (S) PTE Ltd
38 Liu Fang Road, Jurong, Singapore 2262
Tel: (265) 6666
Fax: (261) 7875

USA and Canada

Books International Inc
PO Box 605, Herndon, VA 22070, USA
Tel: (703) 435 7064
Fax: (703) 689 0660

Payment can be made by cheque or credit card (Visa/Mastercard, quoting number and expiry date). Alternatively, a *pro forma* invoice can be sent.

Prepaid orders must include £2.50/US$5.00 to cover postage and packing for one item and £1.25/US$2.50 for each additional item.